J. Volke, F. Liška

Electrochemistry in Organic Synthesis

With 18 Figures and 12 Tables

Springer-Verlag

Berlin Heidelberg New York
London Paris Tokyo
Hong Kong Barcelona Budapest

Dr. Jiří Volke
The J. Heyrovský Institute of Physical Chemistry
Academy of Sciences of the Czech Republic
Dolejškova 3, 18223 Prague 8, Czech Republic

Dr. František Liška
Institute of Chemical Technology
Technická 5, 16000 Prague 6, Czech Republic

ISBN-13: 978-3-642-78701-0   e-ISBN-13: 978-3-642-78699-0
DOI: 10.1007/978-3-642-78699-0

Library of Congress Cataloging-in-Publication Data
Volke, J. (Jiří), 1926- Electrochemistry in organic synthesis / J. Volke, F. Liška.
Includes bibliographical references.

1. Organic compounds - Synthesis. 2. Organic electrochemistry.
I. Liška, F. (František), 1940-. II. Title

Typesetting: Macmillan India Ltd., Bangalore-25
SPIN: 10077106   51/3020 - 5 4 3 2 1 0 - Printed on acid-free paper

Preface

This book has been written as an introduction to the electro-synthesis of organic compounds, in particular for organic chemists. Both authors assume that the knowledge of electro-chemistry of these specialists is rather poor and is usually based only on the remnants of the teaching in the courses on physical and analytical chemistry during their university studies. Even with Czech chemists one cannot expect – as it was in the past – the experience obtained in the courses on polaro-graphy.

This is the reason why it was deemed necessary to write an introductory text to the electrosynthesis of organics both as regards the theoretical and the methodological point of view, i.e. the fundamentals, the experimental setup, the application of various working and reference electrodes, the shape and con-struction of electrolysis cells, the use of suitable protic and aprotic solvents, the experience obtained with various sup-porting electrolytes, the separation and isolation of products, as well as the use of inert gases which prevent the interaction of intermediates and of final products with, for example, oxygen or traces of water. – The second part of the book contains a systematic description of preparative organic electrochemical processes, the interpretation of their mechanisms and several prescriptions for synthesizing characteristical groups of com-pounds.

As a whole the book is not written in an exhaustive way. Its final aim is to inform the organic chemist about the possibil-ities and the limitations of these methods both in synthesis of organic compounds and in the interpretation of mechanisms of organic redox reactions as they appear in the early 1990s.

J. Volke
F. Liška

Prague, March 1994

Contents

1 Introduction

The history of the application of electric current for preparing organic substances [1] had already begun 150 years ago. At that time Faraday in his attempts to oxidize electrolytically the salts of aliphatic acids first discovered the formation of the corresponding alkanes. The actual beginning, however, is considered to be the year 1849 when Kolbe interpreted the above reaction and used it purposefully in the synthesis of alkanes. In 1898 Haber prepared phenylhydroxylamine and aniline selectively by electrolytic reduction of nitrobenzene, he found that phenylhydroxylamine results at less negative potentials and that 4 electrons per molecule of nitrobenzene are consumed in its formation. When the reduction of nitrobenzene was performed at more negative potentials, aniline was prepared with the consumption of 6 electrons. In this way a discovery was made which had a decisive importance for the further development of electrochemistry. It followed from his experiments that the electrode potential is the fundamental factor which determines the value of the Gibbs energy of the electrode process, i.e. of the heterogeneous electron transfer between the electrode and the organic molecule. In this way, theoretical foundations were laid for selective transformations of organic compounds on electrodes. The practical performance of such reactions was made easier by the potentiostat, constructed by Hickling in 1942. This device, when working with a three-electrode system, automatically keeps the potential of the working electrode at the required constant value with a reference electrode. In consequence of this technical innovation a relatively rapid development of organic electrosynthesis was initiated (use of the preceding knowledge of novel organic electrochemistry and of organic polarography was also made) in the mid 1950s and has lasted until now. The development of spectral and, more recently electroanalytical procedures – as well as that of more advanced separation and isolation methods – make it possible to obtain a deeper insight into the structure and reactivity of intermediates which result during the electrode process and react in follow-up chemical and electrochemical processes. Not only the use of potentiostats but also the use of new electrode materials, new materials for diaphragms, non-aqueous (mostly aprotic) organic solvents and novel supporting electrolytes contribute to increasing selectivity of electrochemical processes. Recently, indirect electrochemical procedures have been introduced and are frequently applied for reaching selective oxidations and reductions of organic substrates: in such processes the so-called mediators are used, i.e. electrochemically regenerable redox system. The importance of electrosynthesis, of this "old-new" discipline for the present industrial society may be confirmed by the engineering solution of the construction of highly efficient working cells. However, the development in the 1980s proved that the most suitable field of application is the preparation of relatively small quantities of valuable fine chemicals. The famous method used in the nylon synthesis is more or less an exception. The discipline resulting in this way – electroorganic synthesis which forms an area between organic synthesis and electrochemistry – makes use of the electrolysis in liquid media for preparing organic compounds or for preparing reagents for further application in organic synthesis. It belongs both to laboratory and to industrial procedures.

In its simplest form, an organic preparative reaction can be compared with a chemical reaction which is followed by the isolation of the required product. In the practical performance of both a laboratory preparation or of an industrial process, in particular the first step, i.e. the chemical reaction, is often not completely satisfactory and convenient. The reaction need not necessarily follow the required path and may lead to side reactions and to the formation of side products, isomers and polymers. A particularly inconvenient factor – from the point of view of energetics – is the frequent necessity to work at increased or high temperatures, or sometimes, at high pressure. Practical experience, theoretical considerations but also consulting the literature concerning preparative procedures of organic chemistry published as early as in the first decades of this century, point to the fact that, in oxidations and reductions, electrosynthesis could be more convenient than classical organic synthesis. The required process is initiated by electrical potential applied to the working electrode.

What else attracts synthetic organic chemists to electrochemistry in addition to the possibility of a selective transformation of substrates and to the fact that as a universal reagent an anode is used in oxidations and a cathode in reductions, none of them usually giving side products?

First it is the easy inversion of the polarity [2] of the molecule ("Umpolung"). This always takes place if an electron transfer occurs between the electrode and the substrate in which ions, radicals or ion radicals are formed as the primary intermediates (see below). In classical organic synthesis such change in polarity is achieved by suitable chemical reactions, such as e.g. (1-1) and (1-2)

$$\overset{\delta +}{R-CH_2} \longrightarrow Br \quad \xrightarrow{\ Mg\ } \quad \overset{\delta -}{R-CH_2} \longleftarrow MgBr \qquad (1\text{-}1)$$

$$CH_3-\overset{\delta -}{\underset{\underset{O}{\|}}{C}}-CH_3 \quad \xrightarrow[-HBr]{\ Br_2\ } \quad CH_3-\underset{\underset{O}{\|}}{C}-\overset{\delta +}{CH_2} \longrightarrow Br$$

$$(1\text{-}2)$$

Further it is the frequently high stereoselectivity [3] of chemical reactions to which the electrogenerated particles are liable both at the electrode surface or in its close vicinity. Thus, the proportion of α,β-stereoisomers of unsaturated hydroxyketones resulting in the anodic acetoxylation of dienolacetates (equal to 13.9) is very close to the ratio of isomers resulting in microsomal oxidation (i.e. 14.1) which also occurs at the interface between the solid and the liquid phase; both ratios differ considerably from the ratio between isomers resulting in the chemical oxidation (3.0) by perbenzoic acid, taking place in the bulk of the solution (1-3)

$$(1\text{-}3)$$

Finally, it is also the exceptional reactivity of electrogenerated particles in the vicinity of the electrode before their diffusion back into the bulk of the solution. This case may be exemplified by the alkylation of the carbanion resulting by cathodic reduction of the iminium ion which occurs with a high yield even in strongly acidic media without an antecedent protonation (1-4).

$$(1\text{-}4)$$

2 Experimental Factors and Methods of Investigation of Electroorganic Reactions

2.1 Fundamental Conceptions of Organic Electrochemistry [4]

Electroorganic reactions are often a combination of two processes, the electrode process (E) and the chemical process (C), cf. Fig. 2.1. This sequence may be repeated or the processes E and C may be combined in different ways such as e.g. EEC, ECE, CECE. The sequence may also be CE. The basis of inducing the electrochemical process E is a heterogeneous electron transfer between the electrode and the substrate which, primarily, without subsequent reactions, leads to the formation of a reactive intermediate, i.e. to a radical ion, a cation, an anion or to a radical, depending on the electron configuration of the starting substance (the substrate, the educt) and on the type of the redox processes, i.e. the oxidation or the reduction. Unless the chemical processes are considered, the E type reactions take into account the following possibilities (2-1):

$$A^{2+} \underset{+e}{\overset{-e}{\rightleftharpoons}} A^+ \underset{+e}{\overset{-e}{\rightleftharpoons}} A \underset{-e}{\overset{+e}{\rightleftharpoons}} A^- \underset{-e}{\overset{+e}{\rightleftharpoons}} A^{2-}$$

$$A^- \underset{-e}{\overset{+e}{\rightleftharpoons}} A^\bullet \underset{+e}{\overset{-e}{\rightleftharpoons}} A^+ \tag{2-1}$$

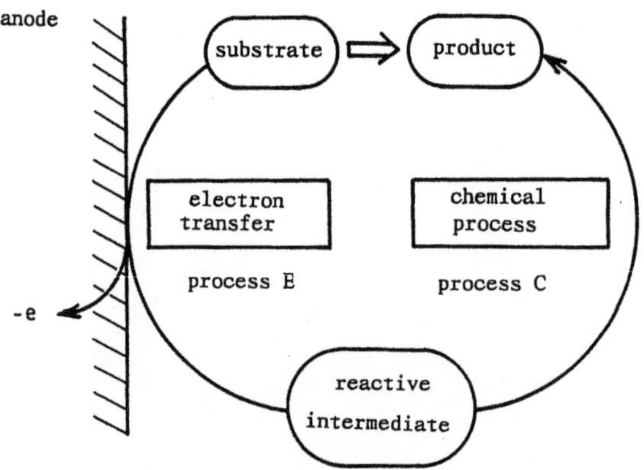

Fig. 2.1. Schematic depiction of an electroorganic oxidation as combination of processes E and C

Keeping all other experimental conditions constant, the fundamental factor which affects an electrochemical process E, is the electrode potential.

The basis of a C type process in electrosynthesis is an aimed orientation of the further course of the reaction and of the reactivity of intermediates formed in the process E in such a way that by corresponding follow-up reactions (additions, substitutions, eliminations, recombination, cleavage, rearrangement etc.) a product of required structure is obtained. This is achieved by the choice of a suitable solvent, supporting electrolyte, electrode material, current density or electrode potential, temperature, pH etc. In this respect electroorganic reactions differ from electroinorganic processes in aqueous solutions which usually end in the process E, although this does not hold true without exception, particularly in coordination compounds.

2.2 Laboratory Electrolysis Cells [5, 6]

For studying electroorganic synthetic reactions on a laboratory scale (1–10 g of the substrate) electrolytic cells are used which possess a rather simple construction. In the simplest case, cf. Fig. 2.2, a glass vessel (a beaker) suffices, in which the working (in case of oxidation the anode) and the auxiliary electrode (the cathode) are placed. In a number of laboratory preparative operations, an electrolytic cell without a diaphragm can be best applied (its construction is depicted in Fig. 2.3). Both electrodes are immersed in the same solution to be electrolyzed, formed by the substrate, the solvent and the supporting electrolyte.

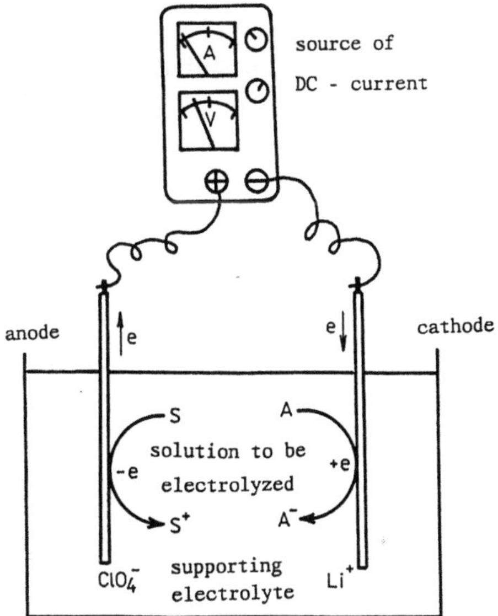

Fig. 2.2. Scheme of electrolysis

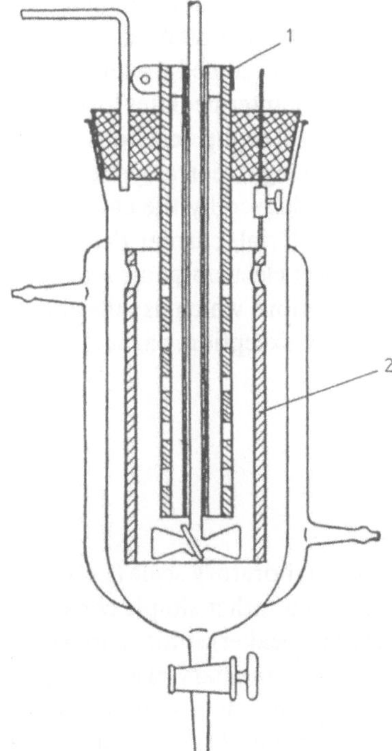

Fig. 2.3. Laboratory electrolytic cell with stirrer but without diaphragm; *1, 2* electrodes

The cell may be equipped with a thermometer, a magnetic stirrer, a tube serving as an inlet for the inert gas and a cooling jacket. The electrodes are connected to a DC-voltage source (0.5–2.5 A; 0–35 V), usually with a built-in voltmeter and an ammeter. The distance between the electrodes varies from 1 to 5 mm in order to make the resistance as small as possible and to achieve current densities at the working electrode ranging from 10 to 100 mA cm^{-2}. Such "undivided" cells without a diaphragm have a low resistance and also the overall voltage (U) is low. In industrial practice this fact leads to a lower energy consumption which is given by the following relationship (2-2),

$$E_g = \frac{U \cdot i \cdot t}{1000} \tag{2-2}$$

where E_g is the energy in kWh, i – the current in A, t – the time in hours, U – the voltage in V.

For a number of electrochemical preparations, in particular reductions, the so-called divided cells must be applied in which the anode and the cathode are divided by a diaphragm which prevents mixing of the electrolytes and thus decreases the cathodic reduction of the product formed by oxidation on the anode and vice versa (cf. Fig. 2.4). The construction of a laboratory electrolysis

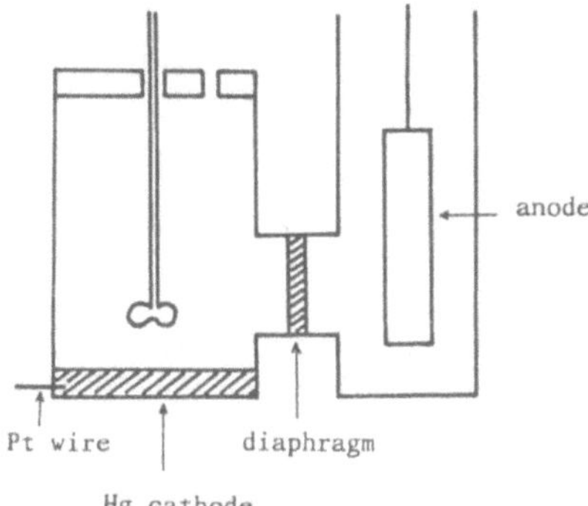

Fig. 2.4. Schematic depiction of an electrochemical cell with a diaphragm

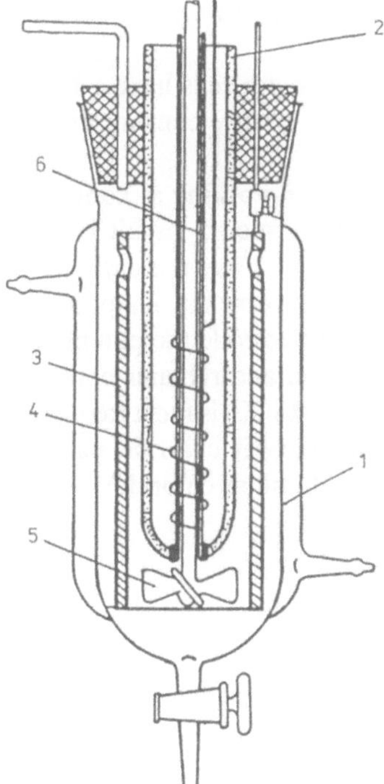

Fig. 2.5. Laboratory electrolyzer with a diaphragm; *1* – glass vessel of the electrolyzer with a water jacket; *2* – ceramic diaphragm; *3* – working electrode; *4* – auxiliary electrode; *5* – stirrer *6* – glass tube

cell with a diaphragm equipped with stirring and cooling is depicted in Fig. 2.5. The electrodes are placed at maximum at a distance of 5 mm from a diaphragm which divides the electrolyte into the anolyte and the catholyte. Ideally the diaphragm should be chemically inert and totally impermeable to the solvent, the educts and the products. However, it should be permeable to the ions. Such an ideal diaphragm has not been produced yet. In practice physical non-selective diaphragms are therefore applied (porous glass, ceramic or plastic

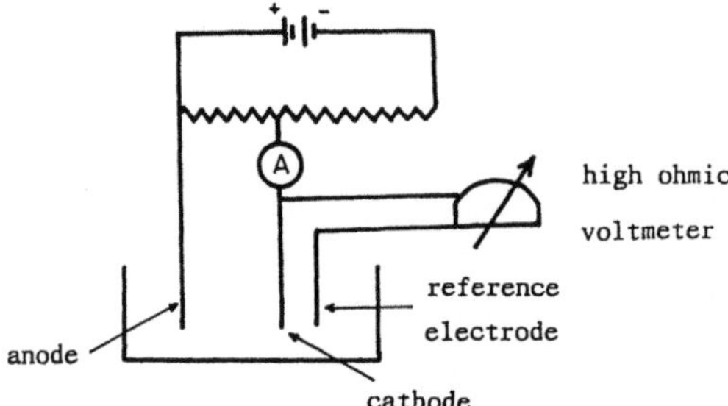

Fig. 2.6. Scheme of the electrolytic cell with controlled voltage and a reference electrode

asbestos, cellophane frits etc.) or semipermeable ionex membranes which are ion-selective, i.e. enable the transfer of only one ion between the anolyte and the catholyte. These are e.g. copolymers of sulfonated polystyrene and divinylbenzene or polymeric perfluorinated membranes of the Nafion type (sulfonated fluoropolymers).

In contrast to the preceding type of cell, the diaphragm cells have a higher resistivity, a higher overall cell voltage and a higher energy consumption in technological processes.

Both the preceding types of cell can be modified for working at controlled potential of the working electrode. A scheme depicting such a circuit is to be seen in Fig. 2.6. The potential of the working electrode (e.g. the cathode) is automatically controlled at the required constant value by means of a potentiostat; this value is measured versus a suitable reference electrode the potential of which vs a normal hydrogen electrode (NHE) is known and remains constant.

In practice the so-called saturated calomel electrode (SCE) is most frequently applied as a reference electrode, i.e. Hg^+/Hg (mercury(I) chloride paste in contact with mercury and a saturated solution of potassium chloride).

2.3 Electrodes [7]

The choice of the working electrode must be carefully considered because its material or the pretreatment or the modification of its surface can fully change the mechanism of the electrode process, the properties of the resulting intermediates, the follow-up reactions and, consequently, also the character of the final product. On its surface the transfer of electrons between the substrate (educt in preparative reactions) and the electrode occurs. The direction of this transfer decides if the reaction is an oxidation or a reduction – i.e. the uptake or loss of the electron by the electrode. This transfer takes place in the electric double layer at the interface between the electrode and the electrolyte; the thickness of the double layer amounts to about 10 nm and the voltage drop therein can reach values as high as 10^7 V cm^{-1}.

The electrode process can be therefore markedly specific and is chiefly affected by the electrode potential, but also by the adsorptive and catalytic properties of its surface. The electropreparative work – in particular long-term work or, perhaps, electrolysis on a technical scale – requires taking into account further electrode properties such as: good electric conductivity as well as mechanical durability, resistivity against chemical and electrochemical influences, in solid electrodes a relatively large surface per unit area and, finally, the ability to catalyze the proceeding reaction. This last requirement is characteristic particularly of organic anodic processes. The auxiliary working electrode is chosen so that a suitable electrode process occurs on its surface whose products interfere only to a very low extent or not at all even if the cathodic and the anodic compartment are not separated; very often a platinum foil, or a platinum wire, or a platinum grid is satisfactory.

Of a given electrode-electrolyte system only a certain, limited range of potentials is characteristic in the region in which oxidation-reduction reaction can occur. This range is sometimes called "the potential window", a term derived in essence from spectroscopy. The limit on the anodic side depends on the electrode material and on the oxidation region of the solvent or, perhaps, on the oxidation of components of the supporting electrolyte. In an analogous way, the attainable cathodic potentials are limited by the reduction potential of the components of the supporting electrolyte or of that the solvent at the given electrode.

In aqueous solutions and, in general, in protic solvents protons may be reduced and hydrogen evolved. For this reason, the potential range of electrodes in the cathodic regions is given by the potential of the H^+/H_2 electrode and its hydrogen overpotential (2-3):

$$2\ H_2O\ +\ 2\ e^-\ \longrightarrow\ H_2\ +\ 2\ OH^- \tag{2-3}$$

If one requires the applicability of the electrode up to distinctly negative potentials, it must possess a low value of the exchange current (i_0) for the reduction of the hydrogen ions. The value of $-\log i_0$ is highest (i.e. the hydrogen evolution is at slowest) with mercury, lead, thallium, manganese and cadmium, the intermediate group is formed by titanium, niobium, tungsten, gold and nickel whereas a high rate of hydrogen ion reduction is characteristic of iridium, rhodium, platinum and palladium. This sequence has been found for aqueous 1 M–H_2SO_4 solutions.

A useful guide for choosing the cathode material is the hydrogen overvoltage of metals which can acquire values up to 1.2 V (vs SCE) with Pb, Hg and Cd whereas in metals used as hydrogenation catalysts (Pt, Ag, Ni, Cu) its values approach zero.

It follows from the preceding paragraphs that the first group is most suitable for carrying out preparative electroreductions owing to its high overvoltage. In essence, best results were obtained with mercury, cadmium and lead electrodes: this follows from older electropreparative publications and – as regards mercury electrodes – from the results of classical (DC-) polarography. A negative property of mercury electrodes – in addition to their state of aggregation and their toxicity – is their easy oxidizability at not very positive (or at negative) potentials, in particular in presence of halogenides, cyanides and further compounds which form poorly soluble or undissociated complexes with mercury cations.

The potential range of electrodes in the anodic region for aqueous solutions is determined by the potential of an O_2/H_2O electrode and by its oxygen overvoltage (2-4)

$$2\ H_2O \longrightarrow O_2 + 4\ H^+ + 4\ e^- \tag{2-4}$$

The body of materials for the choice of anodes is substantially limited with respect to the conditions of oxidation, in particular in preparative electrolysis. The best known material is platinum where, however, the most serious hitch is its high price. This is why in the industrial practice it is replaced by platinized anodes on a well-conducting support from titanium and also by PbO_2 electrodes. Further materials are carbon or graphite. In spite of some of its convenient properties (higher hydrogen overvoltage) gold is not very frequently applied; this is probably due to complications with sealing it into glass.

Noble metal electrodes are not inert at sufficiently positive potentials because in aqueous solutions they form oxide layers on the surface. The stoichiometry of such compounds is relatively not well defined. It seems that in a strongly positive regions the oxide film is composed of chemisorbed oxygen with the nucleation and with the growth of the oxide phase. This holds for Pt, Pd, Rh and Au. Platinum electrodes are especially suitable for one-electron oxidations, i.e. for the primary formations of radicals and radical ions whereas on carbon electrodes, two-electron oxidations occur under otherwise identical conditions; these mechanisms lead to the formation of cations. In polar aprotic solvents, the formation of the chemisorbed layer on a platinum anode is less pronounced and on polished platinum, the highest positive limit can be reached among all electrodes applied in electrochemistry. The condition is the absence of water and for its removal a perfect technique is required which is usually hardly achievable in preparative work. The noble metal electrode surface is polished and renewed by the so-called cycling, i.e. gradually scanning the applied potential in a range from the given value toward the positive side and back to a suitable negative value, followed by a return to the starting potential. In this way a so-called active surface is obtained. When describing this operation, it is not always emphasized that in aqueous solutions a considerable quantity of the electrode material is dissolved. Platinum and gold electrodes, however, are dissolved to a lesser degree than palladium and rhodium. In experiments with a gold electrode it is necessary to take care that potentials applied to it in solutions containing halogenides or cyanides are not too positive; this prevents their oxidation to e.g. tetrachloroaurate anions (2-5):

$$Au + 4\ Cl^- \longrightarrow AuCl_4^- + 3\ e^- \tag{2-5}$$

The carbon electrodes comprise the carbon electrodes proper and the graphite electrodes. The materials used include glass-like carbon which exhibits a good conductivity and a sufficient resistivity versus chemical effects. Its advantage is the low price and high overvoltage both for oxygen and for hydrogen. A further form, spectroscopic graphite, is very porous and for this reason it is impregnated with paraffin or ceresin wax. It can only seldom be applied for electrochemical preparations. The third group is represented by pyrolytic graphite in which the hexagonal rings are parallel to the electrode surface. This form is chemically most resistant and thus protected against the penetration of gases. In connection with the choice of carbon anodes one has to point out that perchlorates – which are so useful with platinum anodes – cannot be recommended here. The most suitable supporting electrolytes are *p*-toluene-sulfonates. The working electrodes in preparative electrochemistry (since the 1980s the impact has been on the production of fine chemicals; for this reason small-scale electrolysis plays the most important role) can be divided into two limiting kinds: in the first case the electrode represents just a sink or a source of electrons – in such a situation the mechanism and the products are independent of the electrode material and the current is controlled by the electrode area. In the other extreme case the material exerts the influence of a catalyst and strong dependence on the electrode material can be observed.

In general, the decisive parameters which control the behaviour of both types of electrode material are as follows: the electrode potential (or current density), the concentration of the species to be electrolyzed, the solvent, the electrolyte, the proton availability, the temperature, the mass transport and the cell design – perhaps also the presence of additives.

The electrodes proper exhibit the following important properties which considerably affect the working out of a new electropreparative method:

physical stability (no abrasion),
chemical stability (e.g. no chemical oxidation),
suitable shape,
rate and products selectivity (influence of electrocatalysis),
low cost and long lifetime,
low toxicity (danger when working with Hg, Cd or Pb).

The relationship between the electrode material and the mechanism of an organic electrode process was investigated for the first time in the mid 1960s. The electron transfer can only occur via the following three ways: a) the reaction takes place via a bond with the electrode surface, b) the reaction occurs without a bond formation and proceeds simply as an electron transfer or, c) in a reduction via adsorbed hydrogen on Pt, Pd or Ni. This type of reduction is very close to catalytic hydrogenation. The hydrogenations occur as follows:

Nickel

ketones → alcohols
aldehydes → alcohols

acetylenes → *cis*-alkenes
olefins → alkanes
unsaturated ketones → ketones
nitriles → amines
Schiff bases → amines
oximes → amines
pyridine → piperidine
cyclohexadiene → cyclohexane
benzene → cyclohexane
sugars → sugar alcohols

Palladium

ketones → alcohols
acetylenes → *cis*-alkenes
nitriles → amines
unsaturated ketones → ketones
unsaturated steroids → steroids
cleavage of benzyloxycarbonyl from peptides

Platinum

ketones → alcohols
ketones → alkanes
butadienes → alkenes
acetylenes → *cis*-alkenes
nitrocompounds → amines
$CF_3COOH \rightarrow CF_3CH_3$

Rhodium

phenols → cyclohexanols

Cobalt, Iron

nitriles → amines

The applicability of an electropreparative method is best characterized by the following definitions:

Current efficiency is the fraction of the total charge passed that is used in the formation of the desired product (the hydrogen evolution is here a competitive reaction).

Material yield is the fraction of the starting material that is converted into the desired product (its value is less than one, the losses are due to by-products, to isolation and to purification). The term space-time yield is less often encountered: it means the weight of product per unit time or unit volume in a given cell.

The reference electrodes do not in essence differ from the reference electrodes known from voltammetric, polarographic (or potentiometric) measurements. In general, they are used in potentiostatically controlled electrolyses with a three electrode circuit.

Evidently, the most frequently used is a calomel electrode with various concentrations of KCl, but also electrodes of the HgO/Hg- and Hg_2SO_4/Hg-type, further the more recently introduced Ag/AgCl electrode (in particular in commercially produced devices), or electrodes of the type of metal electrodes in a solution of their own cations (e.g. Ag in a solution of Ag^+ ions). When working with non-aqueous electrodes the solution in the electrode is e.g. 0.1 M $(C_2H_5)_4NClO_4$; the Ag^+ concentration is equal to 0.01 M. The reference electrode is usually separated from the solution to be investigated by a bridge containing the electrolyte, either the same or at least with one ion in common with the salt present in the reference electrode. The bridge is separated from the electrode as well as from the electrolytic compartment by dense frits or frits supported by agar plugs. This holds for aqueous solutions and the situation does not substantially differ with polar aprotic solvents. In spite of the fact that – in contrast to measuring methods with controlled potential (voltammetry and polarography) – here it is not necessary to know the exact values of potentials, the choice and preparation of a reference electrode for non-polar non-aqueous solvents is a difficult problem. Non-polar solvents (e.g. dichloromethane) lead to ion association (with respect to the low dielectric constant) and to an increase in the resistance of the solution. In this case the reference electrodes are prepared in a solvent miscible with dichloromethane (the condition is that such an electrode is reliable and its potential is not time dependent) or a reference electrode is made which is based on a half-cell in dichloromethane (2-6).

$$\frac{Ag/Ag_3I_4^-}{(C_4H_9)_4\ N^+} \tag{2-6}$$

In preparative electrochemistry our ignorance of the answer is not important to the question if for each non-polar solvent a special electrode is required or if a universal electrode can be constructed which is suitable for all aprotic non-polar solvents.

The main requirement of the experiments in voltammetric or even more in electropreparative experiments is that the potential of the reference electrode remains constant during the whole measurement or during the electrosynthetic procedure. It is not so important that its potential corresponds to theoretical assumptions since this value can be measured directly or by means of a standard with a known reduction or oxidation potential ($E_{1/2}$ or E_p); $E_{1/2}$ or E_p can be determined voltammetrically or polarographically, compared then with the value known from the literature and the result used for correcting the measured value. Such a quasireference electrode may be e.g. a large area Hg/pool electrode in a solution of halogenide ions (Cl^-, Br^- or I^-) or a platinum wire or a foil. The reproducibility is sufficient for electropreparative aims, b i.e. $\pm$ 10–20 mV. The most frequently used quasireference electrode is Ag-wire.

Table 2.1. Standard potentials of reference electrodes $E^{0'} + E_j$ (V at °C)

Reference electrodes	Molarity	$E^{0'} + E_j$ (V at °C)	
		20	25
AgCl/Ag	3.5 M KCl	0.208	0.205
	saturated	0.204	0.199
Hg_2Cl_2/Hg	0.1 M KCl	0.336	0.336
	1.0 M	0.284	0.283
	3.5 M	0.252	0.250
	saturated	0.248	0.244
Hg_2SO_4/Hg	saturated K_2SO_4	0.658 (22 °C)	
Hg/HgO	0.3 M NaOH	0.926	

2.4 Solvents and Supporting Electrolytes [8]

Electrolytic reactions and, consequently, also electropreparative processes occur mostly by a heterogeneous electron transfer between the electrode and the substrate in the solution and are followed by further processes in the liquid phase containing the substrate, the solvent and the supporting electrolyte. The most suitable solvent i.e. water, is only seldom used in organic electrochemistry. When choosing a suitable solvent, not only the solubility of the starting material (the substrate) has to be considered, but also the solubility of the primary and of the final products and, last but not least, the solubility of the supporting electrolyte. Whereas the starting material and the supporting electrolyte must be easily soluble in the given system, it is in some cases advantageous if the product is insoluble and deposits during the electrolysis. The relative permittivity of the solvent has to be larger that 10 if possible: this ensures a suitable dissociation of the supporting electrolyte and the conductivity of the solution. The chosen system, i.e. the solvent and the supporting electrolyte should be inert toward the starting material and the final product; further it must enable the separation of the compound thus formed without difficulties and its oxidation and reduction must be more difficult than that of the substrate.

In oxidations, acetic acid, pyridine, nitromethane, but in particular acetonitrile (AN) and related compounds are used. AN exhibits a low viscosity, adequate volatility, can be relatively easily purified and allows the separation of products. An unfavorable property of AN is its toxicity. Owing to its dielectric constant the solutions of salts even at $0.05 \, mol \, l^{-1}$ concentrations have a satisfactory conductivity. AN is not such a strong base as dimethylformamide (DMF) or dimethylsulfoxide (DMSO); this leads to the fact that in anhydrous AN the radical cations are more stable than in DMF whereas the radical anions have here a much lower lifetime. The applicable potential range is shown in Table 2.2.

Table 2.2. Potential range in acetonitrile under different conditions

Supporting electrolyte	Electrodes working	reference	Potential range (V)
$(C_2H_5)_4NClO_4$	Hg	SCE	0.6 to -2.8
$LiClO_4$	Pt	Ag/0.01 M $AgClO_4$/ 0.1 M $LiClO_4$	2.4 to -3.5
$NaBF_4$	Pt	Ag/0.1 M $AgNO_3$	anodic to 4.0

Table 2.3. Potential ranges in dimethylformamide under different conditions

Supporting electrolyte	Electrodes working	reference	Potential range (V)
$(C_2H_5)_4NClO_4$	Hg	SCE	$+0.5$ to -3.0
$(C_2H_5)_4NClO_4$	Pt	SCE	$+1.6$ -2.1
$(C_4H_9)_4NClO_4$	Hg	SCE	-0.4 -3.0
$(C_4H_9)_4NClO_4$	Pt	SCE	$+1.2$ -2.5

Although a perfect purification of AN is difficult, in the pure state it can be stored very well. Nevertheless it is hygroscopic, light-sensitive and it ages rapidly in experiments and manipulations. Commercial, spectroscopically pure AN can usually be used in electrochemistry directly. Owing to their stability (non-reactivity) other nitriles are generally used in electrochemistry, e.g. propionitrile or benzonitrile which have very similar chemical and electrochemical properties but are more expensive and not so easily available.

Tetrahydrofuran, 1,2-dimethoxyethane, diglyme, pyridine, dimethylformamide, dimethylacetamide and a whole series of aliphatic alcohols (in particular methanol and ethanol) or acetone, appear to be suitable solvents for electro-reductive processes.

The amides are of considerable importance because of their high dielectric constant. They are usually very resistant toward electroreduction but not very suitable for anodic processes. Most frequently used is dimethylformamide which in addition has a low vapor pressure and a negligible toxicity. It is not suitable for anodic reactions. As supporting electrolytes tetrafluoroborates and hexafluorophosphates of tetraalkylammonium cations and alkali metal cations are applied as well as perchlorates and halogenides (cf. Table 2.3).

The limiting process at a platinum anode at about $+1.5$ V is the loss of a single electron from the nitrogen atom of DMF. Traces of water cause a decomposition of DMF (hydrolysis to formic acid and dimethylamine – the same mechanism as in alkaline media). The solvent purified by relatively complicated procedures based on distillation is stable for several weeks if kept in the dark and in a refrigerator. A lower tendency to hydrolysis has been found in *N*-Methylpyrrolidone. *N*-Methylformamide has a very high dielectric constant.

Among the amides of inorganic acids hexamethylphosphortriamide is used which can form stable solutions of electrons. On the cathodic side with a lithium salt as the supporting electrolyte potentials as high as -3.6 V can be reached whereas with a tetraalkylammonium salt only -1.1 V is attainable at Pt. A substantial hazard is the carcinogenity of this compound.

One often uses ethers which have a very wide potential range; their great drawback is the low dielectric constant and, as a result of this, a rather high resistance; moreover, under the influence of light and air they form peroxides. They include the formerly often used toxic dioxane and 1,2-dimethoxyethane. The most important of the ethers is tetrahydrofuran (THF), which is very stable toward reductive agents. The widest applicable potential range has been found with a platinum working electrode and $LiClO_4$ as supporting electrolyte: it is from $+1.8$ to -3.6 V vs Ag/AgI. For the purification of this solvent not only distillation with $LiAlH_4$ is used but the so-called ketyl drying by means of benzophenone where the indicator of the absence of water is the intensively blue radical anion of the ketone. In addition to the above solvents which have been classified according to related chemical structures one must also take into account some structurally different but still important solvents.

Dimethylsulfoxide is at present a very frequently applied polar solvent in which, owing to electron donation, association of molecules occurs. This also plays a positive role in the association with water: at the same rest humidity, in a similar way as in DMF the radical anions are here more stable than in acetonitrile. The experiments can be performed here with most common supporting electrolytes. Pure dimethylsulfoxide for spectral use can be directly applied or it can be purified by vacuum distillation and drying on a molecular sieve. In contrast to the other solvents it is not toxic but it easily and usually rapidly penetrates through the skin and all tissues so that it can possibly transport into the body dissolved toxic agents. The applicable potential range is shown in Table 2.4.

Propylene carbonate is a cyclic ester with a high dielectric constant. It is non-toxic and non-reactive and very easily dissolves organic and inorganic substances. It stabilizes the resulting radical ions. Unfortunately it contains a large quantity of impurities; for this reason it is purified by a multiple fractionated vacuum distillation and the drying is accomplished on molecular sieves. With $(C_4H_9)_4NClO_4$ as supporting electrolyte and at a platinum electrode the potential range is believed to be from $+1.7$ to -1.9 V (vs SCE) and at a mercury electrode from $+0.5$ to -2.5 V. However, recent studies demonstrate that the anodic stability of propylene carbonate is much lower than considered previously.

Table 2.4. Potential ranges in dimethylsulfoxide under different conditions

Supporting electrolyte	Electrodes working	reference	Potential range (V)
$NaClO_4$	Pt	SCE	$+0.7$ to -1.85
$(C_4H_9)_4NClO_4$	Hg	SCE	$+0.4$ -2.7
$(C_4H_9)_4NI$	Hg	SCE	-0.4 -2.85

For anodic oxidations of aromatics, nitromethane ($\epsilon = 36$) is used as solvent. It is relatively unstable and decomposes during storage. The use is limited to special cases. Dichloromethane is a solvent suitable for working at low temperatures (it is very volatile) when it stabilizes radical cations much more than the other solvents. The solvents only rarely applied in organic electrosynthesis are sulpholane, pyridine, nitrobenzene etc. Rich information concerning solvents in electrochemistry is to be found in the monographs by Mann [10] and in the book by Sawyer and Roberts [11].

The above-mentioned knowledge concerning the properties of non-aqueous solvents and electrolytic reactions performed therein in absence of water are made use of quite principally when studying and interpreting the mechanisms of organic electrode processes. When working out new electropreparative methods, a considerable number of authors, particularly organic chemists lacking a deeper electrochemical education, work chiefly empirically. These chemists choose solvents in which the substrates are easily dissolved and – particularly in the early days of preparative electrochemistry – they solved the problem by passing on to a compromise and applying mixed solvents for electrolysis. Usually they took mixtures of water with methanol or ethanol. The choice of the solvent system was therefore at the beginning controlled by the solubility of all substances which participate in the sequence of electrode and chemical reactions, especially that of the educt. In some procedures successful results were obtained in electrolyses with suspensions or still better with emulsions of educts. The most recent step in the development which solves complications connected with the electrolysis of poorly soluble educts is based on the principles of phase transfer catalysis cf. Chapter 3.3

The supporting electrolytes [9] used in organic preparative electrolysis usually differ from supporting electrolytes applied in fundamental electrochemical research, i.e. in voltammetric or polarographic measurements. Even when working with aqueous solutions, buffers are only seldom used and this holds also for cases in which it is known that the required reduction or oxidation mechanism occurs optimally at a given concentration of hydroxonium ions. With respect to the fact that in contrast to voltammetry the electrochemical reaction occurs practically with the whole amount of educt present in the solution, the capacity (and the corresponding concentration) of the buffer should be substantial. For this reason, when working with an aqueous solution one prefers solutions of weak or strong acids, further, for obtaining alkaline media solutions of alkali metal hydroxides, carbonates or acetates, very often, however, salts of strong bases and strong acids. In general, one prefers tetraalkylammonium and lithium cations (the latter group is not recommended for mercury electrodes) as cations of the supporting electrolyte. The reduction potentials of cations become more negative in the following sequence:

$$Na^+ > K^+ > N^+(C_2H_5)_4 > N^+(C_4H_9)_4 > Li^+$$

The choice of suitable supporting electrolytes is especially important in the work with nonaqueous solutions where the electrolytes have to decrease the solution resistance. An example of this property is given in Table 2.5.

Table 2.5. The influence of the anion on the specific resistance of tetra-butylammonium salts in different solvents

Supporting electrolyte	Specific resistance (Ω cm^{-1})		
	acetonitrile	dimethoxyethane	dimethylsulfoxide
$(C_4H_9)_4NClO_4$	37 (0.60)[a]	312 (1.0)[a]	77 (0.60)[a]
$(C_4H_9)_4NBF_4$	31 (1.0)[a]	228 (1.0)[a]	69 (1.0)[a]
$(C_4H_9)_4NBr$	48 (0.60)[a]		106 (0.60)[a]

[a] in brackets: concentration in mol l^{-1}

When applying quarternary ammonium salts as supporting electrolytes, in aprotic media potentials up to about -2.9 V can be reached. This group comprises salts with methyl, ethyl, very often *n*-butyl, hexyl but also phenyl groups; these groups may be also combined in the cation. In the electrosynthetic preparation of adiponitrile according to the process of the Monsanto Company [12] the application of *N,N,N',N'*-tetrabutyl-*N,N'*-diethyl-1,6-hexanediammonium hydrogenphosphate (2-7)

$$
\begin{array}{ccc}
C_4H_9 & & C_4H_9 \\
{}^+| & & {}^+| \\
C_4H_9\!-\!N\!-\!CH_2CH_2CH_2CH_2CH_2CH_2\!-\!N\!-\!C_4H_9 \\
| & & | \\
C_2H_5 & & C_2H_5 \\
& HPO_4^{2-} &
\end{array}
\tag{2-7}
$$

makes possible to decrease the working voltage from 11.65 V to 3.84 V and, consequently, to decrease the energy consumption.

For working in nonaqueous media hexafluorophosphates, hexafluoroborates, perchlorates and *p*-toluenesulfonates are chosen as anions in tetraalkylammonium salts since they are more resistant toward oxidation. The resistance of anions of supporting electrolytes toward oxidation increases in the following sequence:

$$I^- < Br^- < Cl^- < ClO_4^- < BF_4^- < PF_6^-$$

In Table 2.6 voltammetric potential ranges at a platinum electrode in 0.1 M $(C_4H_9)_4NClO_4$ as supporting electrolyte are shown for different organic solvents.

Information about the influence on the oxidation potential and on the reduction potential of the anions and cations of the supporting electrolyte in the same solvent is given in Table 2.7.

It is evident, as follows from the above table – unless a halogenide is necessary for the electrode process proper or for the follow-up reactions – only the last three anions play an important role in anodic processes. Particularly convenient are the tetrafluoroborates and hexafluorophosphates which make possible the achievement of the most positive potentials. Nevertheless, there are

Table 2.6. Potential ranges in common organic solvents (Pt-electrode, 0.1 M $(C_4H_9)_4NClO_4$, SCE)

Solvent	Dielectric constant	Potential range (V)	
tetrahydrofuran	7.6	$+ 1.10$ to $- 2.1$	
methylformiate	8.5	$+ 1.20$	$- 1.60$
methylenechloride	9.08	$+ 1.35$	$- 1.70$
pyridine	12.0	$+ 1.20$	$- 2.10$
acetone	21.0	$+ 1.00$	$- 1.60$
ethanol	24.3	$+ 0.65$	$- 1.20$
benzonitrile	25.5	$+ 1.70$	$- 1.96$
methanol	32.6	$+ 0.70$	$- 1.00$
nitromethane	35.7	$+ 1.15$	$- 1.15$
N,N-dimethylformamide	36.7	$+ 1.30$	$- 2.60$
acetonitrile	37.5	$+ 2.10$	$- 2.30$
N,N-dimethylacetamide	37.8	$+ 1.10$	$- 2.30$
dimethylsulfoxide	46.6	$+ 1.20$	$- 2.70$
propylenecarbonate	64.4	$+ 1.20$	$- 1.50$
1-methyl-2-pyrrolidon	–	$+ 1.10$	$- 1.10$

Table 2.7. Oxidation and reduction potentials of anions and cations in anhydrous acetonitrile

Anion	Oxidation potential (V)	Electrode indicator	reference
CNS^-	$+ 0.55$	Pt	SCE
Cl^-	$+ 1.1$	Pt	SCE
Br^-	$+ 0.70$	Pt	SCE
I^-	$+ 0.30$	Pt	SCE
ClO_4^-	$+ 0.60$	Hg	SCE
ClO_4^-	$+ 2.10$	Pt	SCE
BF_4^-	$+ 2.91$	Pt	$Ag/10^{-2}$ M Ag^+
PF_6^-	$+ 3.02$	Pt	$Ag/10^{-2}$ M Ag^+
Cation	Reduction Potential (V)		
Li^+	$- 1.95$	Hg	SCE
Na^+	$- 1.85$	Hg	SCE
K^+	$- 1.96$	Hg	SCE
Rb^+	$- 1.98$	Hg	SCE
Cs^+	$- 1.97$	Hg	SCE
NH_4^+	$- 1.83$	Hg	SCE
$(C_4H_9)_4N^+$	$- 2.30$	Hg	SCE

two drawbacks: such salts can only be used in completely anhydrous media; much more important is the fact that they are very expensive and this would play an especially negative role in large scale electropreparations. From this point of view, more convenient is the situation when using perchlorates which, however, are somewhat dangerous in purification and in drying where explosions have occurred in several cases. A similar decision must be made between lithium and tetraalkylammonium (mostly tetrabutylammonium) cations. With the latter group distinctly more negative potentials can be attained but a serious drawback is again a higher price and/or a more difficult preparation of salts. It would seem therefore that if reaching only somewhat less negative potential is not disadvantage, lithium salts would be more convenient in preparative electroreductions but in this case their considerable effect on the mechanism of the electrode process and on the composition of products would play a negative role.

A special mention deserves sodium tetraphenylborate and the tetraphenylaluminate which are used as supporting electrolytes when working with ethers as solvents, e.g. with tetrahydrofuran.

2.5 Inert Gases

In most cases both research in laboratories and in small scale and large scale procedures, one has to carry out preparative electrolyses in an inert atmosphere, in particular in the absence of even traces of oxygen. Its presence may substantially change or modify the reduction mechanisms. As an inert gas which is mostly – in contrast to polarographic or voltammetric measurements – passed through the solution during the whole duration of electrolysis, nitrogen, argon and helium may serve, exceptionally also hydrogen, propane, methane or a mixture of hydrogen and nitrogen. Carbon dioxide is not necessarily inert and may be used for electrolytically introducing a carboxylic group. Nitrogen is used chiefly and most frequently and this is due to its low price and easy availability. The oxygen content and its humidity vary and are often high. Moreover its inertness cannot be guaranteed since e.g. even at room temperature it can react with lithium or form coordination compounds with some transition metals and its reactivity at higher temperatures is not negligible even though electrolyses of organic compounds usually do not occur at very high temperatures. This is why with sensitive systems one prefers helium and especially the cheaper argon. The latter contains relatively little water and oxygen and is heavier than air. For this reason, even without stirring or continuous passing through the solution it forms a protective layer above the solution to be electrolyzed.

The last traces of oxygen may be removed from inert gases by passing it through an aqueous solution of Cr^{II} or V^{II} salts which are simultaneously regenerated by amalgamated zinc. In this way even negligible traces of oxygen – down to 1 ppm – can be removed, but the solution is saturated with water which must be – in nonaqueous, aprotic solvents – removed again by means of a drying device with a large capacity. A further method is based on leading the inert gas through a quartz tube filled with copper shavings at a temperature of 450–800 °C. The gas must be cooled down to a temperature necessary in the electrolytic cell. A combination of the BTS catalysts with molecular sieves is more modern. Such a system can be repeatedly regenerated.

For drying, a column filled with solid potassium hydroxide can be also used and for the final drying a column with phosphorus pentoxide dispersed on glass wool in order to prevent clogging of the column or to prevent the formation of inactive channels in the P_2O_5 layer. One can further desiccate with the help of silica gel, anhydrous calcium sulfate, magnesium perchlorate or, perhaps, one of the above-mentioned molecular sieves. The inert gas after removing oxygen cannot be passed through rubber or polyethylene tubing: outer oxygen may easily and rapidly diffuse through it.

All these exacting requirements, however, concerning the removal of water and oxygen from the solutions to be electrolysed are taken into consideration particularly in mechanistic and molecular electrochemistry directed at the interpretation of mechanisms of organic electrode processes.

2.6 Information Obtained by Electroanalytical Methods

2.6.1 Possibilities of Electrochemical Methods [12]

Most experimental data in organic chemistry have been obtained by methods in which the transport of the electroactive substance toward the electrode is controlled only by diffusion, i.e. convection has been excluded. Owing to this situation, the concentration of the starting material at the electrode during electrolysis decreases and that of the products increases. Such methods in which the factors i, E and t (time) play a role are called non-stationary, if the electroactive substrate at the electrode is depleted during electrolysis; its concentration is supplemented by diffusion. According to the shape of the electrode it may be linear, cylindrical or spherical diffusion. The methods in which the electrolytic current is measured as a function of the linearly growing potential (as a function of time) are called voltammetric. A special kind of voltammetry at the dropping mercury electrode (DME) is the classical (Heyrovský) DC-polarography. This method has yielded most data and ideas for the interpretation of the reduction processes. The simplest definition of this technique is as current measurement at the dropping mercury electrode. The definition of IUPAC based on the conceptions of the Prague school, is somewhat different and limits polarography to methods which make use of liquid electrodes the surface of which is periodically or continuously renewed. This holds primarily for the dropping and streaming mercury electrodes. In essence, even with the rotated disk electrode the solution which is in contact with the electrode, could be constantly renewed and thus fulfills the conditions postulated for polarography (in the latter case however, the electrode is not liquid). The transport here is not diffusion controlled.

If the nonfaradaic components of the measured current are neglected i.e., in particular, the so-called charging or capacitive current (corresponding to charging the electrode double layer which can be looked upon as a condenser), one can explain the behaviour in the following way:

The currents caused by transition of the electrons between the two phases, i.e. between the electrode and the solution (the electroactive species which is present in the solution) result in a change of the total number of electrons in the reacting component. Such currents are specified as faradaic currents and are defined by the relationship (2-8):

$$i = n.F.\frac{dN}{dt} \tag{2-8}$$

where n is the number of electrons exchanged by a single particle, F is equal to 96 500 coulombs, dN/dt is the number of electroactive particles which reach the electrode surface per unit of time: in general, the transport of particles toward the electrode is composed of three components: diffusion, migration and convection. The migration component which predominates if no supporting (indifferent) electrolyte is present in the solution and if charged particles undergo the electrode reaction is mainly undesirable. The convectional contribution is given e.g. by the growth of the dropping electrode versus the solution or by rotation of the disk electrode. The diffusion currents are controlled by the rate of diffusion of the electroactive particles to the electrode from the bulk of the solution. In a purely diffusion-controlled current Fick's First Law can be therefore applied (2-9):

$$i = n.F.A.D.\left(\frac{dc}{dt}\right)_{x=0} \tag{2-9}$$

where D is the diffusion coefficient and A the electrode surface area.

By applying Fick's Second Law a relationship (2-10) is obtained which respects also concentration c_0 at the electrode; c_0 is given by potential and the bulk concentration c^* as a function of time and of the distance from the electrode:

$$i = n.F.A.D.\left(\frac{c-c_0}{\sqrt{3/7\pi Dt}}\right) \tag{2-10}$$

By substituting $A = 0.85\,(mt)^{2/3}$ for the electrode (drop) surface and $c_0 = 0$ for the limiting diffusion-controlled current we obtain the so-called Il'kovič equation (2-11) for the instantaneous current:

$$i = 0.732.n.F.c.D^{1/2}.m^{2/3}.t_1^{1/6} \tag{2-11}$$

In this equation m is the flow velocity (flow rate) of mercury from the capillary.

With the mean surface area during the drop time the best known expression for the mean current is (2-12):

$$i = 0.627 \cdot n \cdot F \cdot c^* \cdot D^{1/2} \cdot m^{2/3} \cdot t_1^{1/6} \tag{2-12}$$

t_1 is the drop-time in seconds. Further calculations introduced a correction factor $(1 \pm 3.97 \cdot D^{1/2} \cdot t^{1/6} \cdot m^{-1/3})$ by which the value of i_d has to be multiplied. The positive or the negative sign in this expression depends on the situation if the original species is present in the solution or in the amalgam (the latter case probably need not be considered with organic compounds).

In contrast to the Il'kovič equation which as (2-12) holds for the limiting current of the sigmoidal polarographic wave, the shape of the polarographic wave in the reversible case (in which both thermodynamic reversibility and a rapid establishment of equilibrium are necessary) is derived from the Nernst equation. The final expression is:

$$E = E^0 \pm \frac{RT}{n \cdot F} \cdot \ln \left(\frac{i}{i_d - i} \right) \cdot \sqrt{\frac{D_{ox}}{D_{red}}} \tag{2-13}$$

where E^0 is standard redox potential, and $-$ and $+$ are valid for a cathodic, and an anodic wave, respectively. If both the oxidized and the reduced species are present in solution the following relationship (2-14) holds:

$$E = E^0 \pm \frac{RT}{n \cdot F} \cdot \ln \left(\frac{i - i_{da}}{i_{dk} - i} \right) \cdot \sqrt{\frac{D_{ox}}{D_{red}}} \tag{2-14}$$

The potential $E_{1/2}$ at which $i = \frac{1}{2} i_d$ is called the half-wave potential and represents an important characteristic constant of the given substance which is often tabulated. Figure 2.7 depicts how a change of the electroactive group affects the value of $E_{1/2}$ in a situation where the rest of the molecule remains the same. Under the assumption that $D_{ox} = D_{red}$ one may write $E^0 = E_{1/2}$. Reactions of the type (2-15)

$$Ox \underset{k_{-e}}{\overset{k_{+e}}{\rightleftarrows}} Red \tag{2-15}$$

in which in contrast to the above considerations the heterogeneous rate constants of the electrode transfer are not high enough give polarographic currents which are controlled by the rate of the electron transfer. This reaction type are the so-called irreversible reactions. For the shape of their polarographic

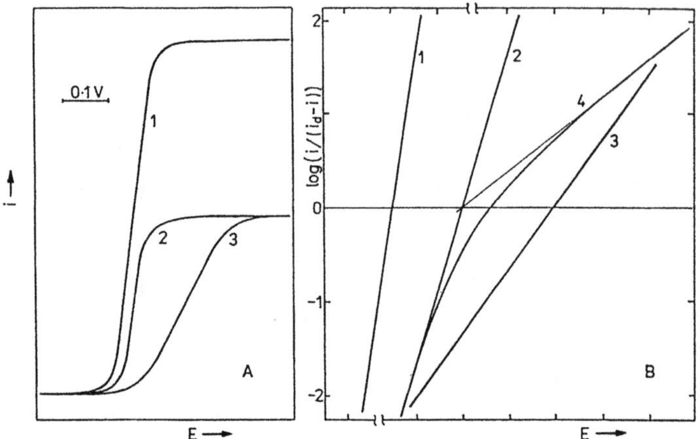

Fig. 2.7. **A**-Shapes of DC polarographic curves for (*1*) two-electron reversible reduction (*2*) one-electron reversible reduction, (*3*) one-electron irreversible reduction; **B** – logarithmic analysis of the preceding curves, i.e. in coordinates log $i/i\alpha - i = f$(E). The plots for *1, 2, 3* are linear but they differ in their slope. Curve *4* holds for the so-called quasireversible case (slight irreversibility): the intersection of the two linear portions enables the determination of the reversible $E_{1/2}$ for a given reaction

wave in the case of a simple unidirectional reduction wave $(k_{-e} \ll k_{+e})$ a relationship has been derived (2-16)

$$\frac{i_{irrev}}{i_{rev} - i_{irrev}} = 0.887 \, k_{+e} \sqrt{\frac{t_1}{D}}$$

$$= 0.887 \, k^0 \sqrt{\frac{t_1}{D}} \, \exp\left[\frac{\alpha.n.F}{R.T} (E-E^0)\right]$$

$$(2\text{-}16)$$

and for the corresponding half-wave potential (2-17)

$$E_{1/2} = E^0 + \frac{2.3 \, RT}{\alpha.n.F} \log\left[0.887 \, k^0 \sqrt{\frac{t_1}{D}}\right] \qquad (2\text{-}17)$$

In this expression k^0 is the standard heterogeneous rate constant which corresponds to the rate at the standard potential E^0 and has a dimension cm.s^{-1}; α is the dimensionless transfer coefficient.

When comparing a reversible and an irreversible sigmoidal DC polarographic wave of a reduction process the following agreements and differences can be observed:

a) in an irreversible process $E_{1/2}$ differs from E^0, is a function of the drop-time t_1 and the anodic and the cathodic wave of the same redox process have different values of $E_{1/2}$,

b) the polarographic wave is more protracted for an irreversible process than for a reversible one.

c) The plot $\log i/(i_d - i)$ vs E does not yield the correct value of n (i.e. the number of electrons transferred in the process) in an irreversible case since it includes also the transfer coefficient α (α must be smaller than 1).

d) The lower limit of the heterogeneous rate constant k^0 is $0.02 \ \mathrm{cm \, s^{-1}}$ for a reversible process which is controlled by diffusion (with lower values the process becomes irreversible). This numerical value holds only for classical DC polarography and different lower limits of k^0 can be found with the various other electrochemical techniques. Consequently, the term "electrochemical reversibility" which has also a kinetic significance differs according to the method applied.

e) Irreversible electrode processes of organic compounds in which protons take part, cannot be analyzed making use of the equation for a reversible system (2-18):

$$E_{1/2} = E^0 - \frac{R.T}{2F} \ln K_1 K_2$$

$$+ \frac{R.T}{2F} \ln\left([H^+] + K_1 [H^+] + K_1 K_2\right) \tag{2-18}$$

This equation holds for the so-called "schema carré" according to Jacq (2-19):

$$
\begin{array}{ccccc}
A & \underset{-e}{\overset{+e}{\rightleftarrows}} & A^- & \underset{-e}{\overset{+e}{\rightleftarrows}} & A^{2-} \\
\Big\updownarrow H^+ & & \Big\updownarrow H^+ & & \Big\updownarrow H^+ \\
AH^+ & \underset{-e}{\overset{+e}{\rightleftarrows}} & AH^{\bullet} & \underset{-e}{\overset{+e}{\rightleftarrows}} & AH^- \\
\Big\updownarrow H^+ & & \Big\updownarrow H^+ & & \Big\updownarrow H^+ \\
AH_2^{2+} & \underset{-e}{\overset{+e}{\rightleftarrows}} & AH^{+\bullet} & \underset{-e}{\overset{+e}{\rightleftarrows}} & AH_2
\end{array}
\tag{2-19}
$$

Equation (2-18) can be simplified according to which of the terms in the bracket prevails (i.e. if $[H^+]^2$, $K_1[H^+]$ or $K_1 K_2$). The plot $E_{1/2} = f(\mathrm{pH})$ is thus composed of three linear portions with slopes 58, 29 and $0 \ \mathrm{mV \, pH^{-1}}$. The points of intersection of these linear portions can be applied for an approximate determination of the corresponding values of pK.

f) In contrast to reversible processes the values of $E_{1/2}$ in irreversible processes may depend on the concentration of the supporting electrolyte.

The second significant method which can be used for obtaining data concerning the mechanisms of electrode processes of organic compounds is voltammetry with a rotated disk electrode (RDE). The rotated disk electrode is usually a planar, circle-shaped area made of conductive material (metal, carbon etc.) mounted into a tube of isolating material (teflon, glass etc.) and conductively connected with the measuring system. The rotation of the disk causes the formation of a solution layer which under the influence of the centrifugal force attains radial velocity. The thickness δ_0 of the layer which horizontally streams along the electrode is equal to

$$\delta_0 = 3\left(\frac{\nu}{\omega}\right)^{1/2} \tag{2-20}$$

here ν is the kinematic viscosity ($|cm^2.s^{-1}|$), and ω is the angular velocity of the disk. If laminar streaming occurs to the electrode the thickness of the diffusion layer δ (2-21) to be applied in the Nernst equation is

$$\delta = 1.61\ D^{1/2}\ \nu^{1/3}\ \omega^{-1/2} \tag{2-21}$$

in which $|D|$ is in $cm^{-2}\ s^{-1}$. The thickness of the diffusion layer is about 10^{-3} cm whereas $\delta_0 = 10^{-2}$ cm. The current for a one-electron process controlled simultaneously by mass transfer and charge transfer is (2-22)

$$i = \frac{F.A.D.c^*}{1.61\ D^{1/3}\ \nu^{1/6}\ \omega^{-1/2} + \dfrac{D}{k}} = \frac{F.A.D.c^*}{\delta + D/k} \tag{2-22}$$

in this relationship A is the area of the electrode (in cm^2), c^* is the concentration of the electroactive substrate (mol l^{-1}) and k is the rate constant of the charge transfer (2-23):

$$k = k^0.\ \exp\left(\frac{-\alpha.F.(E-E^0)}{R.T}\right) \tag{2-23}$$

The potential possesses such a positive value that the counter-reaction can be neglected. The relative values δ and D/k represent two extreme cases in which the rate-controlling process is either the transport of the electroactive substance or the charge transfer. For high values of k the term D/k can be ignored. The expression for the limiting current is then (2-24):

$$i_{lim} = 0.62 \; F.A.c.D^{2/3}.\nu^{-1/6}.\omega^{1/2} \tag{2-24}$$

If, on the other hand, k is very small, the relationship (2-25) is obtained:

$$i_{lim} = F.A.k.c = F.A.c.k^0 \exp\left[\frac{-\alpha.F.(E-E^0)}{R.T}\right] \tag{2-25}$$

In this case the current is controlled by the rate of the charge transfer.

The variation of i with $\omega^{1/2}$ is an important diagnostic criterion for the elucidation of the electrode process. At low potentials (at the foot of the wave) the current is independent of the angular velocity. At high potentials (e.g. at the potential of the limiting current) only the mass transfer controls the current, hence it is directly proportional to $\omega^{1/2}$. At the intermediate potentials the current is first controlled by the mass transfer (small values of $\omega^{1/2}$) and then by charge transfer.

In a fast system without coupled chemical reactions a linear plot $i = f(\omega^{1/2})$ is satisfied regardless of the potential in the $i - E$ curve, i.e. Levich's criterion is fulfilled. In addition to the calculation of the diffusion coefficients or rate constants of the electrode process, this electrode was used for detailed analysis of mechanisms of electrode processes, especially in reactions of the ECE type, where by changing the rate of rotation for example the second step in the electron transfer can be excluded. Thus, the dependence $i_l/\omega^{1/2} = f(\omega^{1/2})$ in the shape of a horizontal direct line reveals that the system is either fast and reversible, controlled by the mass transfer or that the chemical reactions are too fast. If, however, the plot $i_l/\omega^{1/2} = f(\omega^{1/2})$ increases with increasing $\omega^{1/2}$, the system is too slow or an ECE mechanism is operative.

A further refinement of this method is the so-called rotated ring-disk electrode (RRDE). In essence it is composed of two concentric rotated electrodes separated by a ring insulator: hence one of the electrodes has the shape of a ring, the other that of a circular disk. By means of a bipotentiostat two different potentials are applied to these electrodes. At the disk surface the species to be investigated is electrochemically generated, in more complicated processes this species undergoes follow-up reactions; sufficient time is available for these reactions before the products reach the ring electrode at which the primary or the secondary product is electrochemically detected, the latter after a chemical follow-up reaction.

As in the preceding case the qualitative use of these procedures is probably more useful since it enables the detection of products, originating in follow-up processes.

The third important method in the study of electrode mechanisms is cyclic voltammetry. In this technique the potential is continuously varied in the anodic or in the cathodic direction; during this variation anodic or cathodic peaks (corresponding to the oxidation or to the reduction of the studied educt) result on the $i = f(E)$ – curve on a stationary electrode. Starting from the final potential attained in this way the polarization is repeated in the reverse direction, usually back to the starting potential. During this process the peaks corresponding to the electrode process of the primary products (electrochemically generated) or those of further chemical processes are followed. The indicator electrodes can be planar (made of Pt, Au or C), but also disks, surface areas of carbon paste electrode or stationary mercury drops. When working with a planar electrode the regime is that of a semi-infinite planar diffusion; in wire electrodes it is a case of cylindrical diffusion whereas with a mercury drop it is spherical diffusion. With the short duration of electrolysis this complication can be simplified to an almost linear diffusion. The dependence of the applied potential on time ($E = f(t)$) has usually the shape of an isosceles triangle. The rates of the changes of potential with time (scan rates) are in the range from 0.5 V min^{-1} to 500 V s^{-1}. The extreme values are only rarely applied, the middle values between them are very useful. The method usually yields only qualitative results but with respect to the very simple work with the indicator electrode it is most popular in preliminary research of organic systems.

First the peak voltammogram according to Eq. (2-26) is recorded:

$$i_p = k . n^{3/2} . A . D^{1/2} . c^b . v^{1/2} \tag{2-26}$$

i_p is the current in amperes, v is the scan rate in V s^{-1}, k is the Randles-Ševčík constant. It follows from this relationship that the expression $i_p/v^{1/2}/c = $ const. holds for the case of a simple process without subsequent reactions (usually $i_p/v^{1/2}/c = f(v^{1/2})$ is plotted). In a reversible system (with a fast charge transfer) the process exhibits an anodic peak in oxidations with a corresponding cathodic peak in the reverse polarization (back reduction). Both peaks differ by (2-27):

$$\Delta E = E_p^a - E_p^c = 0.058/n \text{ volts} \tag{2-27}$$

The width of the peak is defined by the relation (2-28):

$$E_p - E_{p/2} = 0.058/n \text{ volts} \tag{2-28}$$

Unless a chemical follow-up reaction occurs the ratio of peak heights of the cathodic and the anodic current is approximately equal to one. The measurement, however, is difficult. If the electron transfer is slow the anodic and the cathodic peak are shifted apart along the potential axis; the lower the rate constant of the electrode process the larger the separation.

A practical application of this method can be demonstrated by the case where the educt is oxidized at the potential $(E_p^{ox})_1$ and the resulting species is reduced back at $(E_p^{red})_1$. The process is followed by a chemical reduction in which a secondary oxidized form results which is reduced at a more negative potential $(E_p^{red})_2$; after switching over the direction of polarization and polarizing to more positive potentials the secondary product is reoxidized at the more negative potential $(E_p^{ox})_2$. When reaching more positive potentials the anodic peak at $(E_p^{ox})_1$ appears again and in the next polarization cycle the whole course is repeated. The decisive diagnostic factor in this method is the possibility of changing the rate and the direction of polarization. The validity of the mechanism (2-29)

$$Red_1 \; - \; e \; \rightleftharpoons \; Ox_1$$

$$Ox_1 \; \xrightarrow{k} \; Ox_2 \tag{2-29}$$

$$Ox_2 \; + \; e \; \rightleftharpoons \; Red_2$$

can be proved by changing the rate of polarization. At a very fast polarization the peak Red_1 is substantially increased and the peaks Red_2 and Ox_2 virtually vanish; on the other hand at a very slow recording the peak Ox_1 remains on the voltammogram but the oxidation product is fully transformed to Ox_2 so that Red_1 vanishes and only Red_2 and Ox_2 can be observed (cf. Fig. 2.8).

Further two methods which could be applied are chronopotentiometry and chronoamperometry. As regards these two methods we refer to the literature [14].

The data which in connection with isolated and identified products of electrolysis gives the most valuable information concerning the electrode mechanism, is the measured number n of electrons exchanged per molecule (or particle) in the electrode process. The most valuable method for determining n in a given reaction is coulometry, mostly at controlled potential. The determination of n must be carried out in such media where a sufficiently high electrolytic current is ensured (duration of electrolysis from 10 to 100 minutes; further, over the whole area of the working electrode (usually a large-area electrode) an

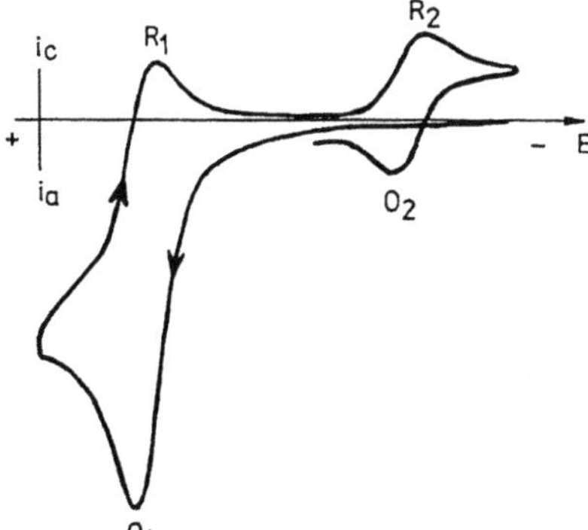

Fig. 2.8. Cyclic voltammogram for a reversible process $R \rightarrow O + ne$ followed by a chemical reaction $O \rightarrow O'$ in which an electroactive product results

accurate constant potential must be kept, the working, auxiliary and reference electrode must be separated but the total resistance must be as low as possible. The measurement is based on Faraday's Law, i.e. on the information that the quantity of electricity required for the transformation of one mole in a one-electron process is equal to $F = 96\,500$ coulombs. In simple cases the electric current decreases during electrolysis according to the relationship (2-30):

$$i = i_0 \cdot \exp(-pt) \tag{2-30}$$

where p is given by Eq. (2-31)

$$p = \frac{D \cdot A}{\delta \cdot \nu} \tag{2-31}$$

i is the electrolytic current at time t, i_0 electrolytic current at $t = 0$, A the area of the working electrode, δ the thickness of the diffusion layer and V the volume of the solution. A larger efficiency of the electrolysis can be achieved by increasing the value of p. An increase of the working area of the electrode and a decrease of the volume lead to this aim; the thickness δ decreases with a higher concentration of the educt and with the efficiency of stirring the solution. The least efficiency has been found with magnetic stirring, where p was calculated to be equal to $1.10^{-3}\,\mathrm{s}^{-1}$; with paddle stirring it is 2.10^{-3}, in screw type stirring 1.10^{-2} and finally in the ultrasonic stirring it reaches $0.5-1.10^{-1}\,\mathrm{s}^{-1}$. The duration of the electrolysis (up to 99.9%) decreases under these conditions from 120 minutes to 1–2.5 minutes.

The change in the value of p as a function of rpm of the stirrer is as about to 2.5 fold of the original value if the number of rotations increases to 1060 rpm from 870 rpm.

The possibility of using coulometry for interpreting electrode processes is closely connected with the choice of a suitable potential which on the one hand corresponds to the potential found on the polarographic or on the voltammetric curve, on the other hand to the potential at which the preparative electrolysis was carried out. However, these potentials may be completely different and also the products may be different. This is why it is necessary to know the electrochemical behaviour of the educt at different potentials. If the potential of the electrolysis is chosen basing on the voltammetric curve, there is a danger of making a serious mistake since the processes may differ if carried out on a micro- or on a macroscale. A simple comparison may be carried out only with uncomplicated reversible processes without adsorption; otherwise the decisive factor for determining the working potential is the result of the voltammetry at a given large-area electrode directly in the coulometric cell or, still better, in the electropreparative cell. Such an electrode has a much larger surface (in comparison to an electroanalytical indicator electrode) and this surface is only imperfectly renewed by stirring, the process occurs therefore at a higher

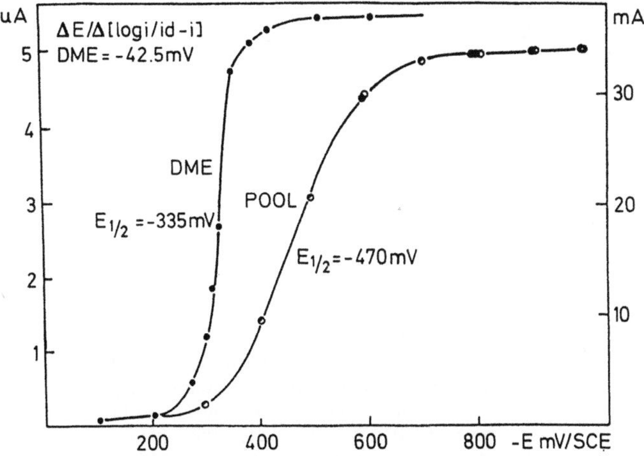

Fig. 2.9. Comparison of $i - E$ curves of monophenylhydrazone of dehydroascorbic acid recorded at a dropping mercury electrode and at a stirred mercury pool electrode; 10^{-3} mol l^{-1} monophenylhydrazone, 0.2 M acetate buffer pH 4.5; 0.93 M KCl; 10% (vol.) dimethylacetamide, R = 35 Ω

concentration and over a substantially longer period of time. This enables parallel chemical reactions to take place as well as adsorption (and its influence on the course of reaction). The half-wave potentials on the $i - E$ curves in classical DC polarography and with a large-area (e.g. mercury pool) electrode may differ by 200 mV or even more as follows from Fig. 2.9. Moreover, such different conditions cause considerable differences in the wave-shapes, in their steepness, number of waves and the mutual ratio of the wave-heights. It can happen that no wave at all appears on the voltammogram with a large-area electrode but nevertheless the electrolytic reaction proceeds (probably not a direct one but a chemical reaction of the educt with the products of the electrode process, e.g. hydrogen, oxygen etc.).

In coulometric measurements, in addition to recording $i - E$ curves (usually in more rapid processes) one continuously follows the concentration of the educt, e.g. by recording polarograms as a function of time. These concentrations are plotted versus charge consumption in coulombs. In simple cases without fast follow-up or parallel reactions, it was found that the dependence $\log i = f(t)$ is linear, the number of transferred electrons is an integer and the results do not depend on the concentration of the educt, on the rate of stirring or on the working potential. Otherwise, the processes are more complicated. Meites [15] worked out a whole system which makes possible the interpretation of a mechanism accompanied e.g. by catalytic processes, by reactions of the educt or of the products with intermediates or products, accompanied by complicating cases of parallel processes, various ECE mechanisms etc. The role of the change in potential [16] in preparative electrolysis and, consequently, also in coulometry can be exemplified (cf. Fig. 2.10) e.g. with a case of reduction of p-fluorobutyrophenone proceeding at a mercury electrode at pH 8.5 (2-32).

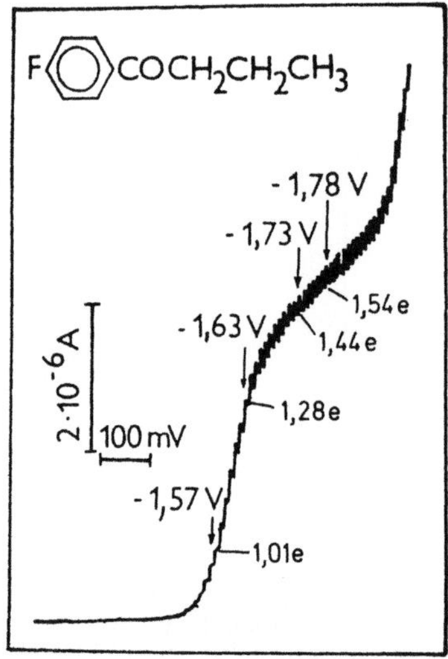

Fig. 2.10. Number of electrons per molecule, n_{app}, measured coulometrically at controlled potential at a stirred mercury pool electrode. The polarographic curve recorded with a DME: $10^{-3}\,\mathrm{mol\,l^{-1}}$ 4-fluoro-butyrophenone, 0.01 M phosphate buffer pH 8.5 (20% vol. ethanol); the numbers on the left: reduction potential at the mercury pool, the numbers on the right demonstrate the value n_{app} obtained in this way

$$\text{reaction with } Hg_2^{2+} \tag{2-32}$$

At a potential which corresponds to the foot of the reduction wave (-1.57 V vs SCE) the value obtained is $n = 1.01$; with increasing negative potentials the experimental value of n also increases, until on the limiting current (at -1.78 V) $n = 1.54$. With the other 4-fluoroarylalkyl ketones this limiting value approaches $n = 2$. The interpretation is as follows: the ketones are primarily reduced in a one-electron step to a radical which can be protonated according to pH; this primary intermediate, however, dimerizes relatively rapidly and easily; the resulting dimer is not reduced further; another possible follow-up reaction of the radical (according to the isolated products) is the formation of a compound with mercury. These reactions compete with the uptake of a second electron under formation of a secondary alcohol. With increasing negative potentials the uptake of the second electron is accelerated and the value of n also grows. At a more negative potential on the limiting current the following relationship (2-33) holds:

$$n_{exp} \longrightarrow (n_1 + n_2), \quad i.e. \ 2 \tag{2-33}$$

2.6.2 Possibilities of Physical Methods

For interpreting electrode processes, non-electrochemical, physical, mainly spectral methods can be applied. These include the UV-VIS region and the IR and NMR spectra and are used chiefly for identifying the products. Such an application is in general not specially affected by the electrochemical procedures preceding their use (in particular if the product has been isolated).

An exception is formed only by the use of especially adapted apparatuses [17], e.g. flow-through cells where it is not necessary to take samples of the electrolyzed solution and to transfer them into cuvettes for spectroscopic measurements; usually such a transfer would be complicated by interference from the presence of oxygen or of water. Even more convenient is a direct combination of electrochemical and spectroscopic methods. In such an arrangement one registers simultaneously both spectral and electrochemical data as a function of the potential during the generation (the spectral absorption and the electrochemical current as a function of applied potential E). Transparent electrodes made from doped SnO_2 and other materials are applied here in most cases.

The most frequent combination of a spectral method with the chemical generation can be encountered in a method particularly important in electrochemistry, i.e. in the proof of existence of radical intermediates or primary products by means of ESR spectroscopy. Only in the formation of very stable radicals they are generated in the electrolytic cell, sealed in the absence of oxygen and transferred to the cavity of the ESR spectrometer. The stability is often substantially increased when measuring at low temperatures (e.g. at $-70\,^{\circ}C$). This is not necessary e.g. with methylviologen blue radicals (their coloration depends on the media, i.e. on the solvent and the electrolyte) which result in a reversible one-electron reduction of methylviologen dications (2-34):

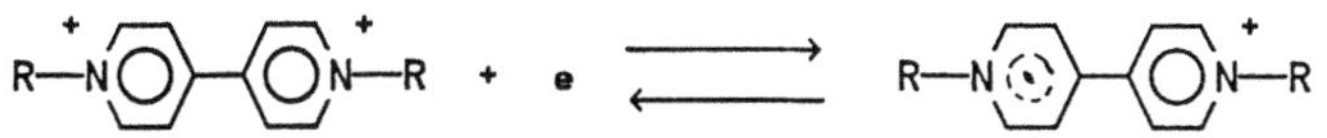

$$R{-}\overset{+}{N}\!\!\bigcirc\!\!-\!\!\bigcirc\!\!\overset{+}{N}{-}R \ + \ e \ \underset{\longleftarrow}{\longrightarrow} \ R{-}\overset{+}{N}\!\!\bigcirc\!\!-\!\!\bigcirc\!\!\overset{+}{N}{-}R \tag{2-34}$$

These radicals are so stable that the authors of this book still have blue samples more than 25 years old kept only in a sealed glass tube at normal temperature in the absence of O_2 and H_2O.

Usually, however, the relatively stable radicals are generated in a special three-electrode cell and from this the solution containing them is transported into the cavity of the ESR spectrometer. Such a generation is called external. In this case the situation for interpreting the spectra is relatively favorable. Better known is the electrolytic generation of the radicals directly in the cavity of the ESR spectrometer; this method was introduced by Maki and Geske [18] for studying the reduction mechanism of nitroaromatics. Here, the complications are caused by the presence of a metallic electrode in the cavity and the difficult deaeration of the solution since the presence of oxygen distorts the ESR spectra (broadening of the lines) and, moreover, oxygen may react chemically with the radicals. For this reason a number of different cells have been developed for generating the radicals directly in the cavity and studying their ESR spectra as a function of time. The main aim is the proof of radical or of radical ion formation as an intermediate of the process. As already mentioned, this intermediate need not be necessarily stable and in the course of time a further particle of non-radical character may result or the radical may generally vanish in a chemical reaction with the media. This is why it is convenient to follow the time-dependence of the ESR spectrum in a state when no voltage is applied to the electrode and no generation of radicals proceeds; in this way the kinetics of their decomposition is investigated (possibly with the corresponding rate constants) but also the relatively frequent formation of a secondary radical whose ESR spectrum may be analyzed and a statement concerning its structure can be made.

The fact that in the electrochemical generation the proof or the existence of the radical by means of ESR was not successful, does still not mean that the radical is primarily not formed but only, if it exists, that it is unstable. In such cases very often indirect methods have proved successful such as the method of electron spin trapping [19]. In this method the primary radical is allowed to react with a compound called the spin trap with formation of a secondary stable radical which can be easily detected, usually by means of ESR. The condition of its formation is the existence of a precursor which exhibits a radical character: consequently, this is the radical to be detected. Convenient spin traps are some nitroso derivates but (since the nitroso compounds are very easily reduced), chiefly the so-called nitrones such as PBN (*N*-benzylidene-tert.-butylamine-*N*-oxide).

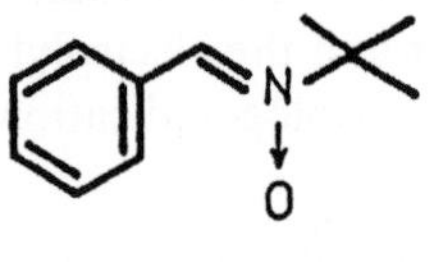

PBN was used for proving the existence of a radical which results in the course of the anodic oxidation of 4-phenyl-1,4-dihydro-1,2,6-trimethyl-3,5-pyridine-dicarbonitrile by deprotonation of the primarily formed radical cation (2-35). The resulting radical adduct is relatively stable and its ESR spectrum can be recorded without difficulties:

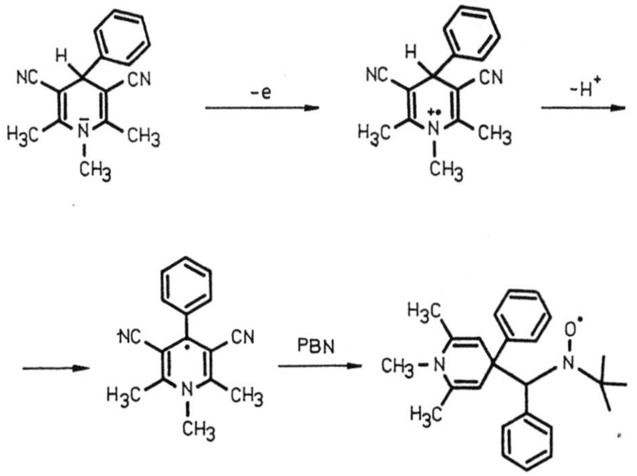

$$(2\text{-}35)$$

The possibility of applying e.g. 2,6-dichloronitrosobenzene as spin trap was attempted in the reduction of pyrylium cations; after the uptake of a single electron neutral unstable radicals result which rapidly dimerize at position 4. A bulky substituent in this position, e.g. a phenyl, lowers the rate of or prevents such a reaction. When comparing different spin traps it was found for substituted pyrylium cations (but not those with a phenyl at 4) that the resulting radical gives an approx. ten-times more intensive ESR signal with 2,6-dichloronitrosobenzene than after a reaction with nitrosodurene or with nitrosomesitylene; on the other hand, the signal of the radical adduct with the above-mentioned PBN is the weakest in the whole series. Substitution merely by a methyl group at 4 of the 1,4-dihydropyridines increases the stability of the primary radical and decreases its reactivity with any spin trap. Even more pronounced is this effect in substitution by a phenyl at 4.

A further indirect physical method for proving the formation of a radical intermediate is based on the electrochemically generated luminescence (ECL) [20]. Luminescence reagents such as e.g. 9,10-diphenylanthracene (DPA) are used here. For the formation of an intermediate radical in the oxidation of substituted 1,4-dihydropyridines a radical cation of 9,10-diphenylanthracene (2-36) is generated at a rotated platinum electrode.

$$\text{DPA} \underset{+e}{\overset{-e}{\rightleftarrows}} \text{DPA}^{+\cdot} \qquad (2\text{-}36)$$

In the same or in a very close potential region a radical cation appears by oxidation of the studied 1,4-dihydropyridine derivative which, however, reacts further and by deprotonation forms the radical RP· (2-37):

$$RPH \xrightarrow{-e} RPH^{+\cdot}$$

$$RPH^{+\cdot} \longrightarrow RP\cdot + H^+ \tag{2-37}$$

By interacting both radical intermediates gradually the excited states ^3DPA and ^1DPA result. By transition from the singlet state to the fundamental state luminescence is emitted ($h\nu$) (2-38) and its appearance is the proof of the existence of the secondary radical $RP\cdot$:

$$RP\cdot + DPA^{+\cdot} \longrightarrow RP^+ + {}^3DPA$$

$$2\ {}^3DPA \longrightarrow {}^1DPA^* + DPA \tag{2-38}$$

$$^1DPA^* \longrightarrow DPA + h\nu$$

The intensity of emitted luminescence I_{ECL} is measured as a function of the electrode potential: usually simultaneously and on the same potential scale two dependencies $I_{ECL} = f(E)$ and $i_{electrochem} = f(E)$ are obtained. The luminescence starts not sooner than at the potential of the anodic oxidation of the luminophore, in this particular case DPA. In addition to the above-mentioned mechanism an indirect mechanism is also possible (2-39) if the luminophore is oxidized at less positive potentials than the oxidation potential of the 1,4-dihydropyridine:

$$DPA^{+\cdot} + RPH \longrightarrow DPA + RPH^{+\cdot}$$

$$DPA + RPH^{+\cdot} \longrightarrow (DPA+RPH)^{+\cdot} \tag{2-39}$$

$$(DPA+RPH)^{+\cdot} \xrightarrow{-H^+,\ -e} {}^3DPA^* + RP\cdot$$

The subsequent sequence of reaction is the same as above. The ECL experiments are not necessarily carried out in the above manner which is called the controlled potential ECL. It is also possible to switch between the oxidation and the reduction potential of the hydrocarbon used as reagent.

2.7 Procedures in Laboratory Electroorganic Synthesis

Every electroorganic synthesis should be preceded by a measurement of the dependence of the electrolytic current on the potential of the working electrode for the given educt (substrate) and for a given composition of the solution to be electrolyzed (solvent, supporting electrolyte, pH, presence of water etc.). In essence this means to carry out the polarographic or the voltammetric research not only in the way characteristic of small indicator electrodes but also at large-area electrodes under conditions very near to those corresponding to normal preparative electrolysis. It was mainly found that the results obtained by electrolysis at small electrodes and on large area electrodes may differ not only in the shape of i–E (or E–i) curves but also in the structure and in the percentual representation of the individual products. For this reason fundamental information is obtained by measuring on these two types of electrodes. Moreover, one must also carry out experiments with different types of electrode materials, at least with platinum, graphite and mercury. Nevertheless it is usually not necessary to perform this whole series of measurements. A lot of information can be found in the literature when the electroactive group is known (in particular as to the possible products). When knowing the fundamentals about the electrolysis of an organic compound one can start the preparative electrolysis proper. One of the two following procedures can be chosen:

a) electrolysis at controlled current density,
b) electrolysis at controlled potential of the working electrode.

The advantage of electrolysis at controlled current density is evident in the case of 1-(2-furyl)ethylacetate methoxylation (2-40) which is carried out with a platinum electrode in methanolic media with $(C_2H_5)_4NClO_4$ as supporting electrolyte and yields almost quantitatively the corresponding 2,5-dimethoxy-2,5-dihydro derivative.

$$\text{(2-40)}$$

Under galvanostatic conditions ($i = $ const.), the potential of the working electrode (anode) adjusts itself in the given system to $E = 1.7$ V (vs Ag/AgCl) and remains at this value during the whole time of electrolysis until the theoretical amount of electricity (here 2 F mol^{-1}) has been consumed. The electrolysis is interrupted after a consumption of 2.05 F mol^{-1} when the potential of the anode is slightly increased to about 1.8 V; this change, however, does not affect the yield and the selectivity of the reaction.

As far as the same methoxylation is carried out at controlled potential of the anode, so with increasing conversion, i.e. with decreasing concentration of the substrate, the current gradually decreases until it approaches zero toward the end of the electrolysis. In this way the duration of the electrolysis is prolonged and the working electrode may be passivated, e.g. by its surface being covered by polymers.

Nevertheless the electrolysis with controlled potential, the so-called potentiostatic procedures, exhibits a number of advantages in many cases. The mechanisms of electrode processes are studied chiefly at controlled potential. In contrast to the galvanostatic procedures the potentiostatic approach ensures a sufficient selectivity of the reaction and is therefore more suitable in those cases

a) where the solution contains several substances which differ in their oxidation and reduction potentials, respectively, and
b) where the substrate undergoes an electrochemical reaction in several steps each of them taking place at a different potential.

The electrolysis with a controlled potential of the working electrode can be performed both on a laboratory or on a large scale; one always works with a single reagent – the electrode whose oxidative or reductive power can be varied in a defined way and continuously over the whole potential range. Thus a selectivity is achieved which mostly cannot be attained with a chemical reagent. It follows from the experience that the difference in potentials between two parallel or subsequent electrochemical processes – one of them should be selectively excluded – should not be smaller than 200 mV.

$$U = E_a + E_c + iR \tag{2-41}$$

The voltage (U) between the two electrodes in the electrolytic cell (cf. Fig. 2.11) is given by the above relationship (2-41) and involves the potential of the anode (E_a) and that of the cathode (E_c); the value of $E_a + E_c$ reaches at maximum several volts; U further involves the considerable value of the voltage drop $i.R$ which, according to the composition of the electrolyte (the kind of solvent, the supporting electrolyte, the arrangement of the cell and of the electrodes) may approach several tens of volts. A very simple way of carrying out the electrolysis at controlled potential is based on measuring the voltage between the working and the reference electrodes and manually, usually with the help of a rheostat, changing the voltage U applied to the working and the auxiliary electrode in

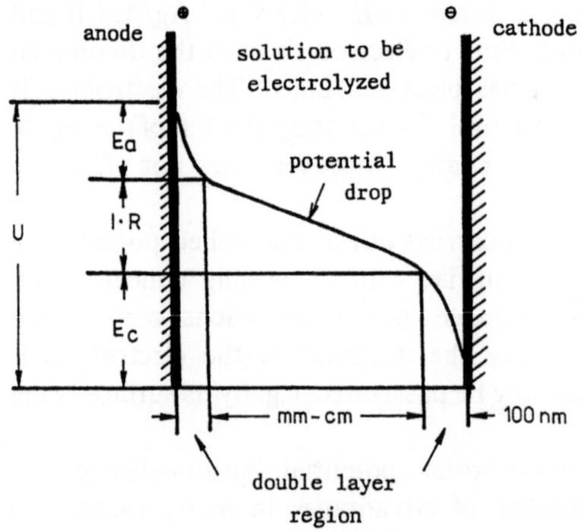

Fig. 2.11. Distribution of voltage in the electrolytic cell; E_a – potential of the anode, E_c – potential of the cathode, i – current in A, R – resistance of the cell in Ω, U – applied voltage between the working and the auxiliary electrode

such a way that the potential of the working electrode is kept on a constant, required value with respect to the reference electrode.

Electrolysis of 9-(2-iodophenyl)acridine on a mercury cathode in ethanolic potassium hydroxide and acetate is an example of preparative electrolysis carried out at constant potential kept manually (2-42).

$$\qquad\qquad (2\text{-}42)$$

The polarogram of 9-(2-iodophenyl)acridine (cf. Fig. 2.12) demonstrates that the reduction occurs in two steps; the first step is characterized by the half-wave potential $E_{1/2} = -1.30$ V (vs SCE) and corresponds to the reduction of the heteroaromatic nucleus with the formation of 9-(2-iodophenyl)dihydroacridine; the second step has $E_{1/2} = -1.62$ V (vs SCE) and corresponds to the splitting off of iodine leading to the formation of 9-phenyldihydroacridine.

As far as in preparative electrochemical reduction the potential of the cathode is kept at values between -1.36 and -1.39 V (vs SCE) by a periodic lowering of the value of the voltage between the electrodes a selective reduction of acridine nucleus occurs and the resulting 9-(2-iodophenyl)dihydroacridine can be isolated by extraction with an almost quantitative yield. In this way the

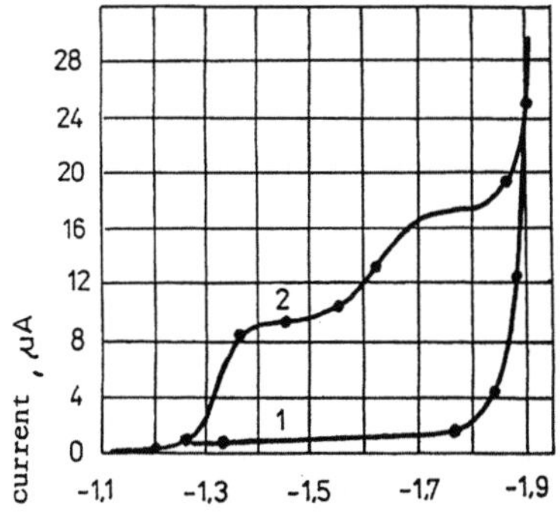

Fig. 2.12. *1* – Polarogram of the supporting electrolyte; the curves are simplified; the oscillations due to the dropping of mercury have been left out. *2* – Polarogram of 2.22 mol.l^{-1} 9-(2-iodophenyl)-acridine in 0.1 M KOH, 0.5 M CH$_3$COOK in 90% ethanol at 25 °C

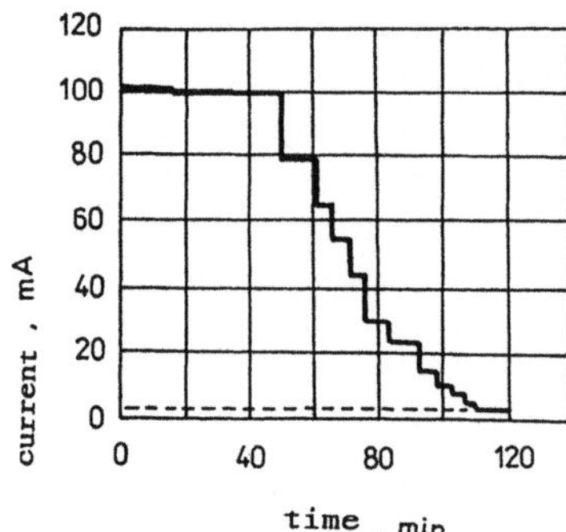

Fig. 2.13. Current–time plot in the case of electrochemical reduction of 9-(2-idophenyl)acridine in 0.1 M KOH with 90% C_2H_2OH carried out at -1.36 to -1.39 V (vs SCE)

formation of the unwanted 9-phenyldihydroacridine is prevented; the reduction up to this degree only takes place at a cathode potential equal to -1.70 (vs SCE).

The current-time dependence of this reduction is depicted in Fig. 2.13. By a graphic integration the consumed charge was determined; in this case it corresponds to the theoretical consumption.

In general, it does not hold, however, that in preparative electrolysis, the same products result as in polarography or voltammetry. In polarography (or in cyclic voltammetry) the transport of the substrate toward the electrode is controlled by diffusion whilst in preparative electrochemical reactions the reaction mixture is intensively stirred and owing to this situation the intermediates in the bulk may undergo completely different chemical follow-up reactions.

The instrument which automatically controls the overall voltage between the working and the auxiliary electrode in such a way that the potential of the working electrode remains constant during the whole duration of electrolysis with respect to the reference electrode, is called a potentiostat. At present, potentiostats are commercially constructed which make it possible to perform potentiostatic electrolysis both on a laboratory scale and on a large scale.

When a coulometer has been introduced into the electric circuit the electric charge (the amount of electricity) can be measured which has been consumed in the electrochemical transformation; this enables the calculation of the current yield (P_n) (the ratio between the charge Q_t calculated from the stoichiometry of the reaction and the charge actually passed Q_{exp}) which is given by the following relationship (2-43)

$$P_n = \frac{Q_t \cdot 100}{Q_{exp}} = \frac{V \cdot n \cdot F}{I \cdot t} \cdot 100 \quad (\%); \qquad (2\text{-}43)$$

here n is the number of electrons transferred between the substrate (educt) and the electrode, V is the mass quantity (mol) transformed at the electrode by passing of Q_t, F is the Faraday constant ($96\,484\ \mathrm{C\,mol^{-1}}$; $26.8\ \mathrm{A\,h\,mol^{-1}}$), I the current passed during the electrolysis (A), t is the duration of electrolysis (h).

The high selectivity of potentiostatic electroorganic synthesis has to be paid for by the prolongation of the electrolysis duration (the current decreases as a result of decreasing concentration of the electroactive substrate). This is why the use of a potentiostat is often circumvented by electrolysis with controlled current. The spontaneous shift of potentials to higher values in the subsequent electrochemical processes is prevented in essence in two ways:

1. The electrolysis is performed with a lower current over the time corresponding to the selective generation of the product resulting in the first electrochemical step.
2. The electrolysis is interrupted after about 90% conversion has been achieved; it is then necessary that the 10% of the starting substrate can be easily separated from the product.

In cases when the product is easily separated from the starting substrate, the solvent and the supporting electrolyte, one can continuously supply the substrate and thus keep "spontaneously" the potential at a value corresponding to a selective transformation of the substrate to the product. This "modus operandi" is often applied in continuous industrial electrosyntheses.

2.8 Research into Mechanisms of Electrode Processes at (chiefly Mercury) Electrodes

In the research of mechanisms of electrode processes of organic compounds at mercury electrodes the first approximation usually consists of the electrolysis at constant potential at the dropping mercury electrode (DME), i.e. with a periodically renewed surface. One works with solutions whose volume is 0.5–5.0 ml and the concentration 10^{-4}–$10^{-3}\ \mathrm{mol\,l^{-1}}$. The identification of the obtained products requires microchemical techniques (or spectral measurements). The next step in electrolysis research is already near to preparative electrosynthesis. The measurement is carried out on a large area stirred electrode – the so-called mercury pool – with the solution having 10^{-3}–$10^{-2}\ \mathrm{mol\,l^{-1}}$ concentrations and the resulting products are isolated by the usual preparative techniques.

In a large number of cases the information obtained in this way by the two methods, agree with each other: the reaction occurs at equal or near potentials, with the same electron consumption and gives rise to products having the same structure. Formally, the same shape appears on polarographic and voltammetric $i - E$ curves as well as on curves recorded with a mercury pool electrode.

In some cases there are considerable differences between the behaviour found on both types of electrodes (i.e. drop vs pool) not only as regards polarographic curves, but also the structure and composition of products. The reasons why only this occurs have been already discussed.

Valuable information concerning the course of the electrode processes can also be obtained from kinetic measurements during electrolysis with a mercury drop. In the simple case of a reversible electron transfer, i.e. for a first-order

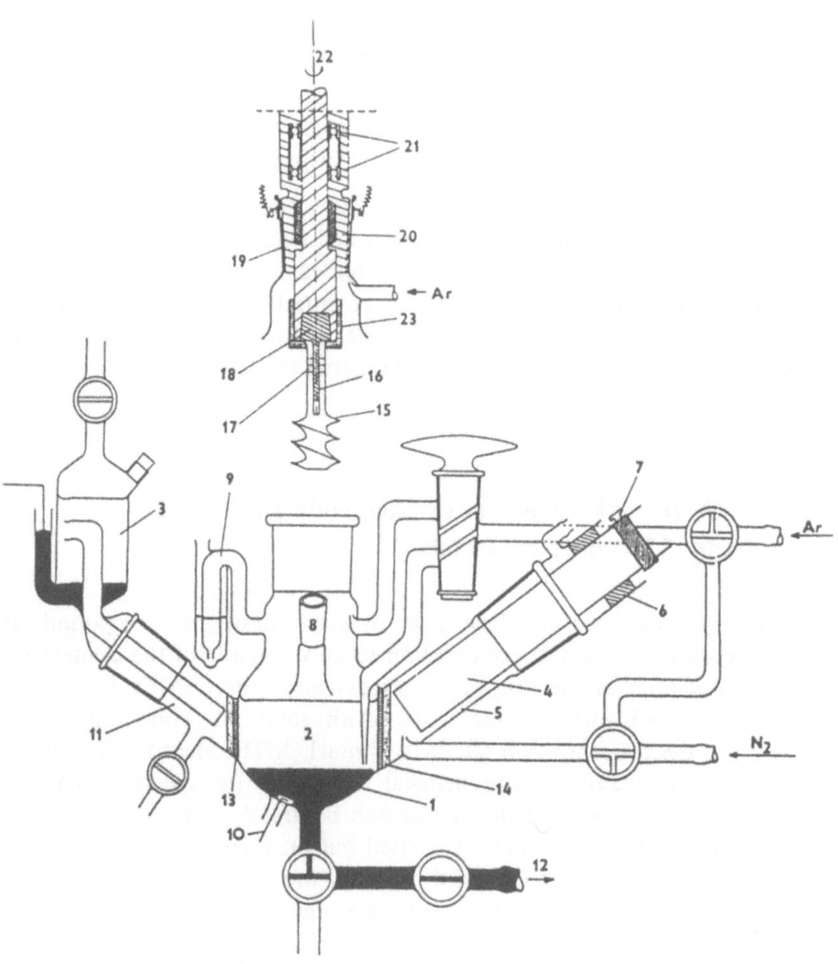

Fig. 2.14. Cell for laboratory preparative electrolysis, for coulometric measurements and recording $i - E$ curves with a large area electrode; *1* – mercury pool electrode, *2* – solution to be electrolyzed, *3* – reference electrode, *4* – platinized titanium auxiliary electrode, *5* – anolyte, *6* – stopper, *7* – contact of the auxiliary electrode, *8* – the dropping mercury electrode, *9* – water seal, *10* – contact to the mercury pool electrode, *11* – salt bridge, *12* – mercury reservoir, *13* – sintered glass disk, *14* – agar bridge, *15* – stirrer, *16* – metallic part of the stirrer, *17* – teflon circles for adjusting the length of the stirrer, *18* – metal part of the stirrer, *19* – teflon foil, *20* – metal shaft of the motor, *21* – simmer ring (for fitting), *22* – motor, *23* – ring nut

reaction, it holds

$$k.t = 2.303 \cdot n \cdot F \cdot v \cdot \log i_0/i_t,$$ (2-44)

where i_0 is the limiting current at time $t = 0$, i_t ... the limiting current at time t, V ... volume in liters, c ... concentrations of the substrate in mol l^{-1}, $k = i_d/c$, the other symbols are well known.

If this relationship is linear one can calculate from the slope $d \log i_t/dt$ (the plot is $\log i = f(t)$) the number of electrons n, transferred in the electrolytic process (2-45):

$$n = -0.4343 \frac{k}{F} \cdot \frac{1}{V} \cdot \frac{dt}{d \log i_t}$$ (2-45)

The diagnostics basing on coulometry of more complicated processes in which not only electron transfers but also competitive and follow-up reactions are operative has been worked out by L. Meites [15].

The differences between $i - E$ curves observed with a dropping mercury electrode and those obtained with a stirred mercury pool can be expressed as a difference in the half-wave potentials (2-46):

$$\left(E_{1/2}\right)_{Hg\ pool} = \left(E_{1/2}\right) + \frac{R.T}{\alpha.n.F} \ln k_e \frac{\delta}{D},$$ (2-46)

δ is the thickness of the diffusion layer, the other symbols are known. For conditions $k_e = 10^{-3}$ cm s^{-1}, $\delta = 10^{-3}$ cm, $D = 10^{-5}$ cm^2 s^{-1}, $\alpha = 0.5$ and $\alpha n = 1$ it can be calculated that the half-wave potential on the stirred mercury pool electrode should be 116 mV more negative. In reversible processes this value is only given by the ratio δ_R/δ_O, i.e. for the reduced and oxidized form and should be very small. In practice the values of $\Delta E_{1/2}$ vary up to 200 mV or even more. The reason why the above equation does not hold quite exactly may be caused by the rate of stirring but also by the adsorption phenomena at the electrode.

For laboratory preparative electrolysis, for coulometric measurements, for recording $i - E$ curves with a mercury pool electrode (or with a DME), or, if necessary for laboratory electrolysis with a non-mercury working electrode, a very practical cell has been developed by Manoušek [21] Fig. 2.14.

References to Chapters 1 and 2

1. Eberson L (1973) In: Baizer MM (ed) Organic electrochemistry. M Dekker, New York, p 470
2. Schäfer HJ (1987) Kontakte (Darmstadt) 17; Schäfer HJ (1987) Kontakte (Darmstadt) 37
3. Shono T (1984) In: Electroorganic chemistry as a new tool in organic synthesis, Springer, Berlin, Heidelberg, New York
4. Baizer MM, Lund H (1983) In: Organic electrochemistry (2nd edn) M Dekker, New York
5. Sawyer DT, Roberts JL, Jr (1974) In: Experimental electrochemistry for chemists. J Wiley, New York
6. Bard AJ, Faulkner LR (1980) In: Electrochemical methods. Fundamentals and applications. J Wiley, New York
   Lund H, Iversen P (1973) In: Organic electrochemistry. In: Baizer MM (ed) M Dekker, New York, p 165
7. Lund H, Iversen P (1973) In: Baizer MM (ed) Organic electrochemistry. M Dekker, New York, p 165
   Couper AM, Pletcher D, Walsh FC (1990) Chem Revs 90: 837
8. Sawyer DT, Roberts JL, Jr (1974) In: Experimental electrochemistry for chemists. J Wiley, New York
   Lund H, Iversen P (1973) In: Organic electrochemistry. M Dekker, New York, p 165
9. Sawyer DT, Roberts JL, Jr (1974) In: Experimental electrochemistry for chemists. J Wiley, New York, p 187
10. Mann CK (1969) Nonaqueous solvents for electrochemical use. In: Bard AJ (ed) Electroanalytical chemistry. M Dekker, New York, p 69
   Mann CK, Barnes K (1970) In: Electrochemical reactions in nonaqueous systems. M Dekker, New York
11. Sawyer DT, Roberts JL, Jr (1974) In: Experimental electrochemistry for chemists. J Wiley, New York
12. Danly DE (1984) J Electrochem Soc 131: 435c
13. Vlček AA, Volke J, Pospíšil L, Kalvoda R (1986) In: Physical methods of chemistry. vol 2. J Wiley, New York, p 797
14. Bard AJ, Faulkner LR (1980) In: Electrochemical methods. Fundamentals and applications. J Wiley, New York
15. Meites L (1960) In: Weissberger A (ed) Techniques of organic chemistry, Part 4, vol 1. J Wiley, New York
16. Volke J, Kejharová-Ryvolová A, Manoušek O, Wasilevska L (1971) J Electroanal Chem 32: 445
17. Krejčík M, Daněk M, Hartl F (1992) J Electroanal Chem in press
18. Kastening B (1972) In: Zuman P, Meites L, Kolthoff IM (ed) Progress in Polarography, vol 3, J. Wiley, New York, p 195
19. Klíma J, Ludvík J, Volke J, Křikova M, Skála V, Kuthan J (1984) J Electroanal Chem 161: 205
   Klíma J, Volke J, Urban J (1991) Electrochim Acta 36: 73
20. Ludvík J, Volke J, Pragst F (1986) J Electroanal Chem 215: 179
   Pragst F, Kaltofen B, Volke J, Kuthan J (1981) J Electroanal Chem 119: 301
21. Manoušek O, Volke J, Hlavatý J (1980) Electrochim Acta 25: 515

3 Reactions of Organic Compounds at Electrodes

Reactive intermediates (radicals, radical ions and ions) which result by an electron transfer may be expected in the region of redox potentials of corresponding electrophores (electroactive atoms or groups of atoms in organic substrates). The precise values of these potentials in actual compounds are always stated for a given electrode, a given supporting electrolyte, a given solvent and are measured versus a suitable reference electrode, most frequently a SCE. An informative, very approximate idea about the potential ranges in which the electrochemical redox transformations of different types of substrates occur is shown in Fig. 3.1. Electrochemical reactions of organic compounds whose redox potentials lie in close vicinity to potentials attainable by electrochemical methods are very difficult to perform in a direct way. In particular in oxidative processes performed at high positive potentials a number of side reactions may proceed (oxidation of the solvent or of the supporting electrolyte) which lower the selectivity of the product formation and hence also the material and current

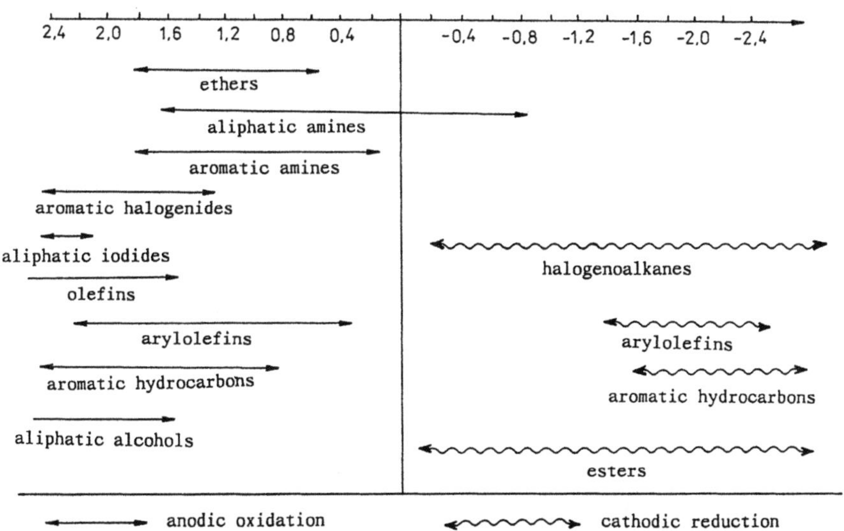

Fig. 3.1. Potential range of oxidation and reduction of some electrophors (vs SCE)

yield, or deactivate the working electrodes by forming polymeric films on the electrode surface.

Much more specific and selective both as regards the educts and the products or components present in the solution are the indirect electrochemical reactions in which the so-called mediator – an electrochemically regenerable redox reagent – undergoes mostly a homogeneous redox reaction with the organic substrate and in a follow-up process is transformed by a heterogeneous electron transfer at the working electrode back to its active form.

This is why the electrochemical transformations are also divided into direct and indirect electrochemical procedures where particularly the latter group has been extensively developed in recent years.

3.1 Direct Anodic Oxidations

3.1.1 Oxidation of Saturated Hydrocarbons

In alkane molecules both the C–H and C–C bonds can be oxidized; in both cases carbocations are formed which undergo further reactions – rearrangements, proton eliminations and reactions with nucleophiles (3-1):

$$
H\!-\!\overset{|}{\underset{|}{C}}\!-\!\overset{|}{\underset{|}{C}}\!-\!H \xrightarrow{-2e}
\begin{cases}
\longrightarrow H\!-\!\overset{|}{\underset{|}{C}}\!-\!\overset{|}{\underset{|}{C}}{}^{(+)} + H^+ & \text{rearrangements} \\[2ex]
 & \text{deprotonations} \\[2ex]
\longrightarrow 2\ H\!-\!\overset{|}{\underset{|}{C}}{}^{(+)} & \text{reaction with} \\
 & \text{nucleophiles}
\end{cases}
$$

$$(3\text{-}1)$$

Direct anodic oxidations can only be performed with alkanes the oxidation potentials of which are lower than about 3.0 V. For this reason the direct anodic oxidations of alkanes are carried out in solvents resistant toward oxidation (acetonitrile, trifluoro-acetic acid, methylenechloride, sulfolan, propylene carbonate etc.) and as supporting electrolytes tetraalkylammonium salts with difficultly oxidizable anions are used ($F^{(-)}$, $ClO_4^{(-)}$, $PF_6^{(-)}$, $BF_4^{(-)}$). The oxidation of CH_3 and CH_2 groups, however, takes place at potentials above 3.4 V ($E_{1/2}$ vs SCE), that of CH groups in the potential range from 3.0 to 3.4 V.

An exceptional position among the alkanes is assumed by adamantane which is oxidized at less positive potentials. In dependence on potential and on composition of the electrolyte, monoacetamido-, diacetamido- and hydroxyadamantanes can be obtained [1, 2], (3-2):

$$(3\text{-}2)$$

Strong mineral acids, such as e.g. fluorosulfonic acid, stabilize electrogenerated carbocations and thus lower the oxidative potentials of alkanes from the values 3.00 V, 3.01 V and 3.40 V ($E_{1/2}$, NEt$_4$BF$_4$, CH$_3$CN) to the values 1.8 V, 1.68 V and 1.64 V (Pd/H$_2$) for 2-methylbutane, 2-methylpentane and octane, respectively. Cyclohexane was in this way – in a mixture of acetic and fluorosulfonic acids at an anode potential 1.85 V (vs SCE) – transformed to 1-acetyl-2-methylcyclopentene which results in a 35% yield [3], (3-3).

$$CH_3COOH + FSO_3H \rightleftharpoons CH_3CO^{(+)} FSO_3^{(-)} + H_2O$$

$$(3\text{-}3)$$

The higher reactivity of tertiary C–H bonds in comparison with CH$_2$ bonds is also demonstrated in the selectivity of oxidation e.g. of decaline to the corresponding decalylacetates [4], (3-4):

$$(3\text{-}4)$$

The oxidation potentials are lowered with increasing inner strain in cycloalkanes. In their oxidation the electrons are transferred to the anode prevalently from the most "strained" C–C bonds (in general from the places of lowest ionization potentials). As an example, in the electrochemical oxidation of tetramethylcyclopropane a cleavage of the C(2)–C(3) bond takes place whilst in the "chemical" solvolysis the proton attacks the sterically least substituted carbon atom [5], (3-5):

$$\text{(3-5)}$$

In the anodic oxidation of tricyclo[4,1,0,0^{2,7}]heptane ($E_{1/2}$ = 1.50 V vs SCE) in methanol a cleavage of the inner bond occurs; the final product of oxidation is

Table 3.1. Oxidation potentials of selected unsaturated compounds

Compound	$E_{1/2}$ (E_p)	Reference electrode	Supporting electrolyte	Solvent	Electrode
C_6H_{13}–CH=CH$_2$	2.8	SCE	LiClO$_4$	CH$_3$CN	Pt
C_5H_{11}–CH=CH–CH$_3$	2.3	SCE	LiClO$_4$	CH$_3$CN	Pt
$(C_2H_5)_2$C=CH$_2$	2.17	SCE	LiClO$_4$	CH$_3$CN	Pt
	2.14	SCE	LiClO$_4$	CH$_3$CN	Pt
	2.02	SCE	LiClO$_4$	CH$_3$CN	Pt
	1.54	SCE	LiClO$_4$	CH$_3$CN	Pt
	1.36	SCE	LiClO$_4$	CH$_3$CN	Pt
X=H	1.54	SCE	LiClO$_4$	CH$_3$CN	Pt
X=COOC$_2$H$_5$	1.85	SCE	LiClO$_4$	CH$_3$CN	Pt
X=CN	1.99	SCE	LiClO$_4$	CH$_3$CN	Pt
CH$_2$=CH–O–C$_2$H$_5$	1.72	Ag/Ag$^+$	NaClO$_4$	CH$_3$OH	Pt
O–C$_2$H$_5$	1.28 (1.63)	Ag/Ag$^+$	NaClO$_4$	CH$_3$OH	Pt
O–COCH$_3$	1.93	SCE	LiClO$_4$	CH$_3$CN	Pt

dimethylacetal of 2-cyclohexenecarbaldehyde owing to a rearrangement of the intermediary carbocation [6], (3-6):

$$(3\text{-}6)$$

3.1.2 Oxidation of Unsaturated Compounds

The oxidation potentials of alkenes and their derivatives are less positive than those compared to alkanes and their values depend on the structure of such an unsaturated compound. It is generally so that the electron-donating substituents (phenyl, alkyl, amino, alkoxy) lower the values of oxidation potentials; on the other hand, the electron-accepting substituents (COR, COOR, CN, NO_2, CF_3, F) increase these values and can change the range attainable in anodic oxidation (potential window). The values of polarographic and voltammetric oxidation potentials ($E_{1/2}$ and E) of selected types of unsaturated compounds are demonstrated in Table 3.1 [7]. In a direct anodic oxidation of a double bond, radical cations result which, depending on the structure of the unsaturated compound and on the supporting electrolyte, give rise to products by oxidative addition of nucleophiles (Nu), by dimer formation, by rearrangement and allyl substitution (3-7). The supporting electrolytes are salts whose anion is oxidized at more positive potentials than the double bond (BF_4^-, TsO^-, PF_6^-, ClO_4^-).

$$(3\text{-}7)$$

In the anodic oxidation of cyclohexene performed in methanol or in acetic acid, derivatives of allyl substitution result (3-methoxy- and 3-acetoxycyclohexenes) accompanied by products of oxidative addition of nucleophiles and/or of rearrangement [8], (3-8).

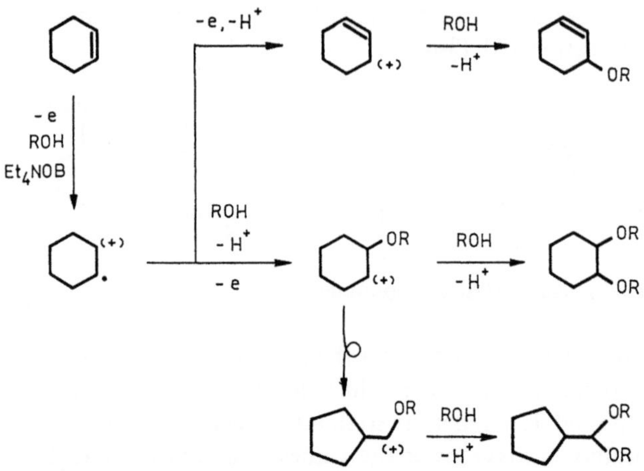

$$(3\text{-}8)$$

The oxidation of cyclohexene carried out in acetonitrile and in presence of a cation exchange polymer (polyvinylbenzenesulfonic acid) – which stabilizes the carbocation resulting by addition of acetonitrile – gives 3-acetamidocyclohexene [9] with a current yield of 63%; in absence of the cation exchange polymer the yield is only 17% (3-9):

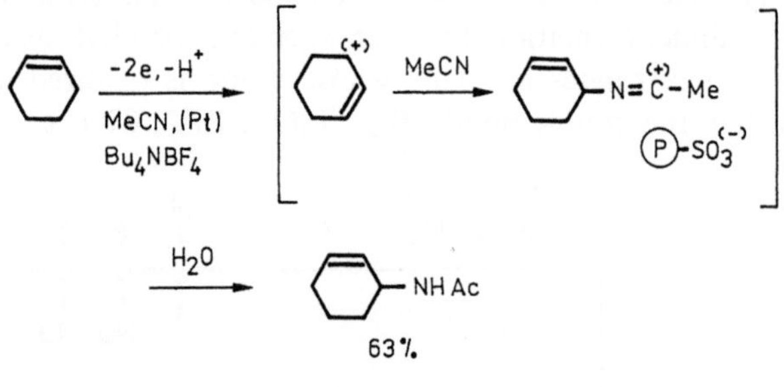

$$(3\text{-}9)$$

63%.

It is interesting to compare the stereochemical course of the electrochemical and chemical acetoxylation of 3-methylcyclohexene. Whereas in the anodic oxidation the thermodynamically less stable *cis* derivative prevails (*cis/trans* = 2.83) the chemical, so-called Kharasch-Sosnovsky reaction results in the thermodynamically more stable *trans* isomer (*cis/trans* = 0.23), (3-10):

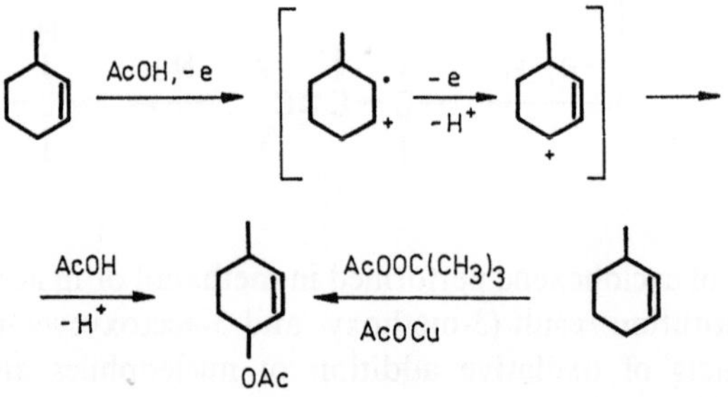

$$(3\text{-}10)$$

The different stereoselectivity of the anodic oxidation is explained by stereo-selective adsorption of the cyclohexene carbocation on the anode surface (3-11).

$$(3\text{-}11)$$

The structure of tetrahydropyrane which results in the anodic oxidation of 3,7-dimethyl-6-octen-1-ol points to the intramolecular mechanism of the alkyl substitution [10], (3-12).

$$(3\text{-}12)$$

As far as the oxidation potential of the supporting electrolyte or that of the solvent is comparable with the oxidation potential of the alkene or is smaller, an addition on the double bond occurs of radicals or of cations resulting in the oxidation of anions of the supporting electrolyte or in the oxidation of the solvent. The main product of cyclohexene oxidation carried out in acetic acid and in presence of tetraethylammonium halogenides are acetates of vicinal halogenhydrines accompanied by products of allyl substitution [11], (3-13):

$$(3\text{-}13)$$

(2-Cyclohexenyl)-acetate [11]

In a 100-ml cylindrical cell equipped with a reflux condenser, a thermometer, and two carbon-rod electrodes (diameter, 0.8 cm) were placed in a solution of 0.10 mol of cyclohexene, 60 g (1.0 mol) of acetic acid, and 5.15 g (0.017 mol) of tetra-ethylammonium p-toluenesulfonate as a supporting electrolyte. This solution magnetically stirred and kept at around room temperature by cooling externally with water, was electrolysed under constant current between 0.1 and 0.2 A until 2 Fmol^{-1} of electricity was passed (current density 32 mA cm^{-2}, anode potential 2.0–2.25 V vs SCE. After the electrolysis was complete, most of the acetic acid was evaporated under reduced pressure. The residue was neutralized with sodium bicarbonate solution and extracted with ether, after drying with anhydrous magnesium sulfate, the ether was removed by distillation. (2-Cyclohexenyl)-acetate was isolated by fractional distillation in 55% yield, bp 64–66°C (2.1 kPa).

The above reaction was ingeniously made use of in the so-called "chlorolactonization" which proceeds with unsaturated acids (3-14); by an oxidation carried out at 1.14V the chloronium ion is generated which is added onto the double bond jointly with the "inner" carboxylate anion under formation of a bicyclic halogenolactone. If perchlorate is used as supporting electrolyte the double bond is primarily oxidized; in absence of the solvent anion only the inner carboxylate anion participates in the addition and by deprotonation of the resulting carbocation a mixture of bicyclic unsaturated lactones results [12], (3-14).

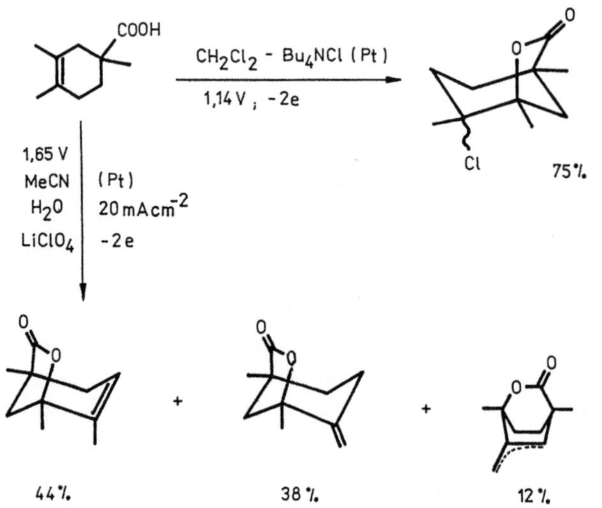

$$(3\text{-}14)$$

The arylalkenes also undergo the oxidative addition of nucleophiles. Thus, in the oxidation of 5-phenyl-4-pentenoic acid the double bond conjugated with the aromatic nucleus is oxidized at a lower potential than the carboxylate anion; by an intramolecular addition of the carboxylate anion jointly with the addition of the "external" methoxyl anion 4-(phenylmethoxymethyl)-4-butanolide results [13], (3-15).

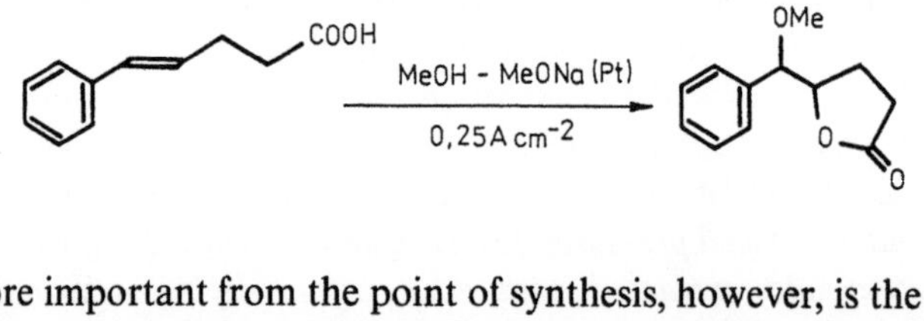

$$(3\text{-}15)$$

More important from the point of synthesis, however, is the oxidative dimer formation of arylolefins which, depending on the composition of the electrolyte, may lead to products of different structure [14, 15], (3-16):

$$(3\text{-}16)$$

The conjugated dienes are oxidized much more easily than simple alkenes. In solutions of methanol or of acetic acid, products of 1,4- and to a lesser degree of 1,2-oxidative addition of nucleophiles are formed (3-17).

$$(3\text{-}17)$$

R = Me   50,6 %   6,1 %

Ac   45,0 %   6,0 %

In case of acyclic dienes dimerization may also take place (3-18).

$$(3\text{-}18)$$

The oxidative 1,4-addition of acetic acid to 3-allyl-2-methyl-1,3-cyclopentadiene was made use of in the synthesis of allethrolone [16], (3-19):

$$(3\text{-}19)$$

The oxidative potentials of dienes with isolated double bonds understandably do not much differ from alkenes. The structure of norbornadiene, however, enables a transannular interaction between the double bonds which makes itself evident by lowering the oxidation potential in comparison with norbornene (cf. Table 3.1) on the one hand, and by increasing the oxidation potential if an electron-acceptor group is introduced at position 2 (cf. Table 3.1) on the other hand. The later situation also affects the composition of products obtained in the anodic methoxylation (3-20).

$$(3\text{-}20)$$

Vinyl ethers and vinyl acetates represent a group of compounds containing electron-donating substituents which lower the oxidation potentials of the double bonds to values easily accessible to anodic oxidation (cf. Table 3.1). Depending on the anode material and on the conditions of electrolysis acetals or 1,4-diketones result which are products of anodic addition of nucleophiles or of dimerization of radical cations (3-21 and 3-22).

$$(3\text{-}21)$$

$$(3-22)$$

From the point of view of synthesis, more meaningful is the oxidation of enolacetates which are more easily accessible. By oxidation, α-acetoxycarbonyl compounds and/or conjugated ketones are formed. Their qualitative ratio depends on the structure of the substrate and on the CH_3COO^- concentration in the electrolyte [19], (3-23):

$$(3-23)$$

$(C_2H_5)_4NOTs$	R = H	3%	37%
	R = CH$_3$	90%	0%
CH_3COOK	R = H	0%	82%
	R = CH$_3$	24%	59%

The anodic α-methoxylation or α-acetoxylation was ingeniously made use in organic synthesis e.g. to 1,2-shift [20], (3-24)

$$(3-24)$$

or to a 1,4-shift of the carbonyl group (3-25):

$$(3\text{-}25)$$

In contrast to the oxidation of enolethers and enolacetates, mixtures of isomeric methoxylated enamines result in the oxidation of enamines [22], (3-26).

$$(3\text{-}26)$$

From the same point of view the anodic oxidation of β-enamino-ketones or esters in which a dimerization of the resulting radical cations proceeds is therefore more significant; the transiently resulting 1,4-dications undergo an intramolecular aminolysis, deprotonation and splitting off benzylamine under formation of a pyrrol nucleus [23], (3-27).

$$(3\text{-}27)$$

3.1.3 Oxidation of Alcohols and Esters

Alcohols, 1,3- and higher 1,n-diols and ethers are not suitable substrates for direct anodic oxidations: the reason are their relatively high oxidation potentials ($E_{1/2}$ vs ferrocene/ferrocenium; CH_3OH, $N(C_4H_9)_4BF_4$: methanol 2.73 V, ethanol 2.61 V, 2-propanol 2.50 V) [24]. The direct oxidation of alcohols is therefore carried out in absence of solvents or in acetonitrile and presence of perchlorates or tetrafluoroborates as supporting electrolytes.

Depending on the structures of the alcohol and on the chosen electrolytic system the electrooxidation of alcohols may lead to the formation of aldehydes, acetals, acids and esters. The course of the direct anodic oxidation of alcohols is interpreted by a mechanism which assumes an intermediary formation of hydroxyalkyl and alkoxyl radicals [25], (3-28). In neutral media the primary process is the electron transfer

$$R-CH_2-\underline{\overline{O}}H \xrightarrow{-e} RCH_2-\underline{O}-H \xrightarrow{-H^+} R-\overset{\bullet}{C}H-OH \xrightarrow[RCH_2OH]{-e,-H^+} R-CH\overset{OH}{\underset{OCH_2R}{<}} \xrightarrow[-H_2O]{RCH_2OH} RCH(OCH_2R)_2$$

$$R-CH_2-\underline{\overline{O}}| \xrightarrow{-e} R-CH_2-\underline{\overline{O}}\bullet \xrightarrow{-e,-H^+} R-CH=O$$

(3-28)

from the free electron pair of the oxygen atom in the hydroxyl group to the anode. The deprotonation of the resulting oxonium radical is the rate determining step of the reaction. The hydroxyalkyl radical formed in this way gives a carbocation in the further oxidation which goes over to a semiacetal in a reaction with alcohol; the semiacetal may yield an aldehyde or an acetal. Alkaline media support the formation of aldehydes. The alkoxide anion is relatively easily oxidized to an alkoxyl radical which is in a further one-electron oxidation transformed to the aldehyde. Both mechanisms participate in the overall oxidation and their ratio depends on the structure of the alcohol and on the reaction conditions (solvent, basicity of the solution). Thus by electrooxidation of 1-butanol in presence of lithium tetrafluoroborate butanal can be prepared in 77% yield. If the oxidation is performed in the presence of hydrochloric acid the main reaction product is 2-chlorobutanal dibutylacetal [26], (3-29).

(3-29)

A preparative significance can be ascribed to the anodic oxidation of 2-butin-1,4-diol to acetylene dicarboxylic acid which is carried out in an aqueous solution of sulfuric acid [27], (3-30):

$$\text{(reaction scheme)} \qquad HOOC-\!\!\equiv\!\!-COOH \qquad (3\text{-}30)$$

$$69 - 76\ \%.$$

In general, one came to the conclusion, however, that a direct oxidation of alcohols to the corresponding aldehydes, ketones, acids or esters is not a very suitable and selective synthetic method and for this reason the indirect electrochemical oxidation is preferred (cf. Sect. 3.3).

On the other hand, the anodic oxidation of *vic*-glycols, *vic*-glycolethers, *vic*-aminoalcohols *vic*-aminoethers and epoxides occurs very easily in methanolic solutions containing $(C_2H_5)_4NTsO$. The reaction products are the corresponding carbonyl compounds [28], (3-31).

$$\underset{OR}{\overset{R^2}{R^1-C}}-\underset{\underset{(NR_2,\ SR)}{OR}}{\overset{R^3}{C}}-R^4 \quad\xrightarrow{-2e\ /\ MeOH}\quad R^1-CO-R^2\ +\ R^3-CO-R^4$$

$$(3\text{-}31)$$

$$R = H,\ Me$$

The ease of such an oxidation is explained by the participation of the vicinal hydroxyl or alkoxyl group in the primary electron transfer from the free electron pair of the atom O or N or S to the anode.

The reaction has been developed for the synthesis of symmetric ketones from methyl methoxyacetate (3-32)

$$CH_3O\!-\!\!\overset{OCH_3}{\underset{O}{\diagup}}\ \xrightarrow{RMgX}\ CH_3O\!-\!\!\overset{R}{\underset{OH}{\diagdown_R}}\ \xrightarrow{-2e}\ R-CO-R \qquad (3\text{-}32)$$

$$89\ \%.$$

$$R = C_3H_7$$
$$C_4H_9$$
$$cyklo-C_6H_{11}$$

and to the synthesis of assymmetric ketones from α-diethylaminoketones [29], (3-33)

$$\underset{O}{\overset{NEt_2}{R\diagdown\!\!\diagup R}}\ \xrightarrow{R'MgX}\ \underset{HO\quad R'}{\overset{NEt_2}{R\diagdown\!\!\diagup R}}\ \xrightarrow{-2e}\ \underset{O}{R\diagdown\!\!\diagup R'}$$

$$R = H, Et;\quad R' = i\text{-}Pr,\ Bu \qquad\qquad 52-72\ \%.$$

$$(3\text{-}33)$$

or from dialkylketones [30], (3-34):

$$(3\text{-}34)$$

65 %.

The primary products of the anodic oxidation of enolethers and enolacetate may also undergo a further oxidative cleavage at the anode under formation of ketoesters [31], (3-35)

$$(3\text{-}35)$$

74 – 94 %.

R = Me, Ac

The same types of compounds result in the anodic oxidation of acyloins [32], (3-36):

$$(3\text{-}36)$$

Methyl-2-methyl-6-oxoheptanoate [31]

Electrolysis was carried out in a H-type two compartment cell (100 ml). The anode compartment, fitted with a drying tube (CaCl$_2$), a thermometer, and a magnetic stirrer was divided from the cathode by 1.8 cm diameter glass-frits plate (No. 56). Two platinum electrodes (3 cm^2) were placed parallel to each other 3 cm apart.

A solution of 85 mg (0.5 nmol) of 2-hydroxy-2-methyl-cyclohexanone and 500 mg LiClO₄ in dry methanol (19 ml) was charged into the anode compartment. Into the cathode compartment a solution of 200 mg of LiClO₄ in dry methanol (16 ml) was charged. The mixture was electrolyzed under a constant applied voltage of 20 V at a current density 5.6–17.5 mA cm⁻² at 30–34 °C. After 7.5 F.mol⁻¹ of electricity was passed, the mixture was concentrated, and the residue was taken up in benzene-ethylacetate (1:1). The extract was washed with brine, dried (Na₂SO₄) and concentrated. The crude product was chromatographed (SiO₂, hexane-ethylacetate, 10:1) to give methyl-2-methyl-6-oxoheptanoate in 94% yield, bp 110–112 °C (2.5 kPa).

Simple aliphatic ethers are assumed to be oxidized in the same way as the alcohols. In their direct oxidation in methanolic solutions and in presence of electrolytes (sodium methoxide, $(C_2H_5)_4NTsO$, ammonium nitrate) α-methoxylated ethers (3-37) are formed in a relatively low yield [32]:

$$R-CH_2-O-R' \xrightarrow{-e} R-CH_2-\underset{}{\overset{(+)\bullet}{O}}-R' \xrightarrow[-H^{(+)}]{-e} R-\overset{(+)}{C}H-OR' \longrightarrow$$

$$\xrightarrow[-H^{(+)}]{MeOH} R-CH\overset{\displaystyle ,OMe}{\underset{\displaystyle `OR'}{}} \tag{3-37}$$

In this way tetrahydrofuran was transformed to 2-methoxytetrahydrofuran; due to changes in the electrolyte composition, however, it undergoes a 4 electron oxidation with the formation of 4-butanolide [33], (3-38):

$$\tag{3-38}$$

3.1.4 Oxidation of Organic Compounds of Sulfur and Selenium

The anodic oxidation of thiols and dialkyl or diaryl sulfides makes it possible to prepare sulfur compounds at different degrees of oxidation in dependence on potential, electrode material (carbon, platinum, stainless steel) and composition of the electrolyte (3-39).

$$
\begin{array}{ccc}
\overset{(-II)}{R-SH} & \xrightarrow{\ -4e\ } & \overset{(II)}{R-\underset{\underset{O}{\|}}{S}-OR} \qquad\qquad R-SO_2O-OSO_2-R \\[2mm]
\searrow{\scriptstyle -H^{+},-e} & & \uparrow \\[2mm]
\overset{(-I)}{R-S-S-R} & \xrightarrow{\ -4e\ } & \overset{(IV)}{R-SO_2O^{(-)}} \\[2mm]
\nearrow{\scriptstyle -R^{+},-e} & & \\[2mm]
\overset{(-II)}{R-S-R} \xrightarrow{\ -2e\ } \overset{(0)}{R-\underset{\underset{O}{\|}}{S}-R} & \xrightarrow{\ -2e\ } & R-\overset{\overset{O}{\|}}{\underset{\underset{O}{\|}}{S}}{}^{(II)}-R
\end{array}
$$

$$\tag{3-39}$$

In the following schemes the particular electrochemical conditions of thiophenol oxidation to diphenyl disulfide are demonstrated and to esters of benzenesulfonic acid [34, 35], (3-40):

$$
\underset{95\,\%}{Ph-\underset{\underset{O}{\|}}{S}-OR} \xleftarrow[\ AcONa,(Pt)\ ]{\ AcOH-ROH\ } PhSH \xrightarrow[\ KH_2PO_4,(Pt)\ ]{\ MeOH-H_2O-KOH\ } \underset{100\,\%}{Ph-S-S-Ph}
$$

$$\tag{3-40}$$

Further the oxidation of diphenyl sulfide to diphenyl sulfoxide, to diphenyl-sulfone and to diphenyl-4-(phenylmercaptophenyl)sulphonium is described [36, 37], (3-41):

$$
Ph-\overset{\overset{O}{\|}}{\underset{\underset{O}{\|}}{S}}-Ph \xleftarrow[\ 1,30V\,;(Pt)\ ]{\ AcOH-H_2O\ } Ph-S-Ph \xrightarrow[\ Cl^-,\,Br^-,\,SO_4^{2-};\,0,94V\ ]{\ AcOH-H_2O\,(Pt)\ } \underset{96\,\%}{Ph-\underset{\underset{O}{\|}}{S}-Ph}
$$

$$
\Big\downarrow{\ MeCN-LiClO_4\ (Pt)\ }
$$

$$
\underset{71\,\%}{Ph-S-\!\!\langle\bigcirc\rangle\!\!-\overset{(+)}{S}Ph_2\ ClO_4^{(-)}}
$$

$$\tag{3-41}$$

Finally the oxidation of diphenyl disulfide to esters of benzenesulfonic acids, phenylbenzenethiosulfonate and to sodium benzenesulfonate are depicted [35, 38, 39], (3-42):

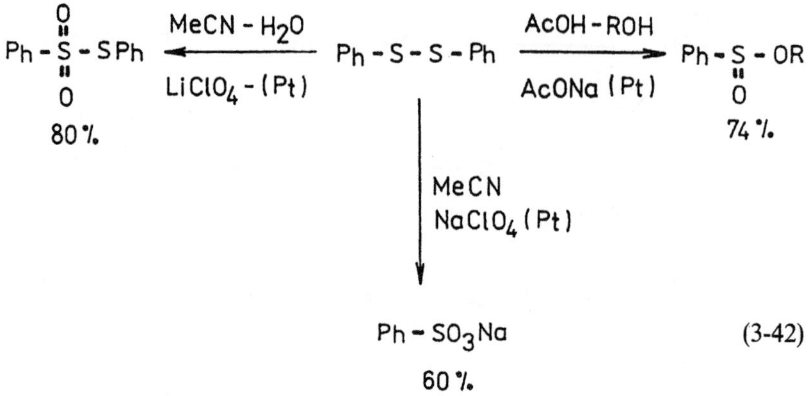

$$\text{Ph-}\overset{\overset{\displaystyle O}{\|}}{\underset{\underset{\displaystyle O}{\|}}{S}}\text{-SPh} \xleftarrow[\text{LiClO}_4\text{-(Pt)}]{\text{MeCN-H}_2\text{O}} \text{Ph-S-S-Ph} \xrightarrow[\text{AcONa (Pt)}]{\text{AcOH-ROH}} \text{Ph-}\overset{\overset{\displaystyle O}{\|}}{\underset{\underset{\displaystyle O}{\|}}{S}}\text{-OR}$$

80 %.   74 %.

$$\downarrow \begin{array}{l}\text{MeCN}\\\text{NaClO}_4\text{ (Pt)}\end{array}$$

$$\text{Ph-SO}_3\text{Na} \qquad\qquad (3\text{-}42)$$

60 %.

One of the methods for the formation of an S–N bond is the preparation of sulfenamides by the electrochemical "oxidative condensation" of thiols and disulfides with primary or secondary amines [40], (3-43):

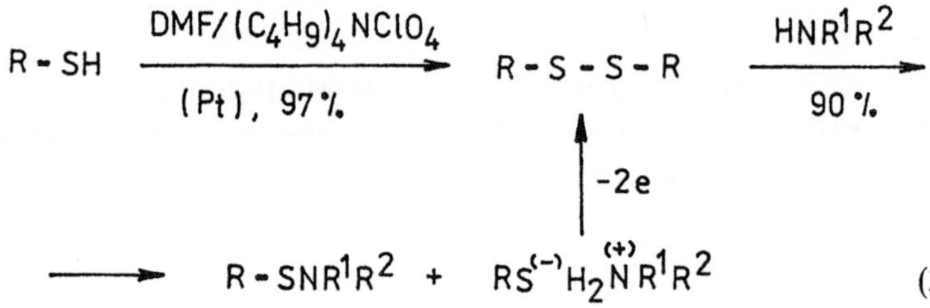

$$R\text{-SH} \xrightarrow[\text{(Pt), 97 \%.}]{\text{DMF/(C}_4\text{H}_9)_4\text{NClO}_4} R\text{-S-S-R} \xrightarrow[\text{90 \%.}]{\text{HNR}^1\text{R}^2}$$

$$\Big\uparrow {-2e}$$

$$\xrightarrow{\quad} R\text{-SNR}^1\text{R}^2 + RS^{(-)}H_2\overset{(+)}{N}R^1R^2 \qquad (3\text{-}43)$$

The fact that in the oxidation of sulfides a sulfenium radical results which may undergo an attack by a nucleophilic reagent "in situ" was made use of for an indirect oxidation of secondary alcohols to ketones [41], (3-44) (cf. also p. 133):

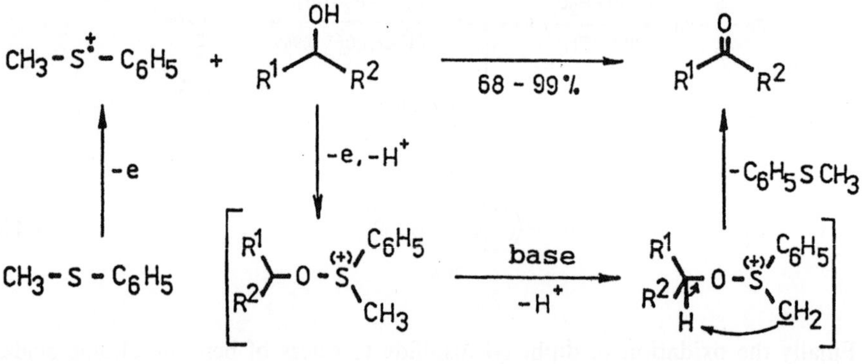

$$(3\text{-}44)$$

The sulfenium radical generated from diphenyl sulfide in an analogous way is used to the "activation" of the hydroxyl in alcohols in S_N2 substitutions (3-45).

$$(3\text{-}45)$$

The anodic cleavage of the C–S bond is applied for removing the dithioacetal protecting group from the molecules of 1,3-dithianes which is carried out in neutral media [42], (3-46). This electrochemical method is a successful competitor of the chemical procedure based on the oxidation of dithianes by Hg^{2+} salts.

$$(3\text{-}46)$$

In recent years the derivatives of benzeneselenic acids ($Ph\text{–}Se\text{–}X$, $X = Cl$, Br, $OCOCH_3$, NR_2) have played an important role in the functionalization of alkenes. Chemically they are prepared by oxidation of diphenyldiselenide, but they are unstable and require a special technique. The electrochemical selenation method is based on the anodic oxidation of a catalytic amount of diphenyldiselenide to the reactive intermediate of benzeneselenic acid which "in situ" adds on to the double bond of the unsaturated compound. The selenide formed in this way is further oxidized on the anode to a selenoxide which by a *syn*-elimination goes over to a hydroxy- or to an alkoxyalkene and the reactive intermediate of benzeneselenic acid [43], (3-47):

$$(3\text{-}47)$$

On estimating this reaction from the point of view of the redox transformation of the organic substrate it can be considered as an indirect electrochemical oxidation in which the benzeneselenic acid plays the role of a mediator (cf. p. 122).

3.1.5 Oxidation of Halogen Derivatives and Oxidative Halogenation of Organic Compounds

Within the range of electrochemically accessible potentials lie only the oxidation potentials of iodo- ($E_{1/2}$ = 1.9–2.1 V vs Ag/Ag$^+$) and bromoalkanes (E_p = 2.5–2.8 V vs Ag/Ag$^+$). By direct anodic oxidation an electron transfer occurs here from the non-bonding orbital of the halogen to the anode; by the following cleavage of the C–Hal bond carbocations are formed which may undergo a Wagner-Meerwein transformation. The original and the newly resulting carbocations react with nucleophiles present in the solution, most frequently with acetonitrile, under formation of N-alkylacetamides [44], (3-48).

$$(3\text{-}48)$$

The above electroacetamidation cannot be performed with fluoro- or chloroalkanes. An example of decreasing oxidizability in halogenoalkanes (RI > RBr $\gg$ RCl > RF) is the electroacetamidation of 1-halogenoadamantanes. Whilst in the oxidation of 1-bromo- and 1-iodoadamantane the carbon-halogen bond is split, in the electrolysis of 1-fluoro- and 1-chloro-adamantane [1, 45] the adamantane nucleus is oxidized at position 3 (3-49):

$$(3\text{-}49)$$

89 %.

X = Cl, 91 %.
X = F, 65 %.

The C–I bond in the aryliodides is not split after the oxidation of the iodine atom; the transient cation radical reacts in a S_E type reaction with a further iodobenzene and in a subsequent oxidation 4-iodophenyliodonium perchlorate is obtained [46], (3-50):

$$(3\text{-}50)$$

The formation of the iodonium radical was ingeniously made use of in the regioselective electrochloration of steroids [47]. The hydroxyl in position 3 is "occupied" by the 3-iodobenzoyl group. The iodonium radical formed in the electrooxidation oxidizes the chloride anion with the formation of the chloro-iodophenyl radical which is located just in such a position that the chlorine radical Cl· set free reaches the hydrogen in position 9 (3-51):

$$(3\text{-}51)$$

The group of oxidative halogenations of organic compounds is based on the electrooxidation of halogenides. The halonium ions formed in this way react with organic substrates in addition or substitution reactions.

In the first case an addition of the halonium ion onto the double bond of the unsaturated compound takes place; in dependence on the kind of the nucleophilic solvent (CH_3CN, CH_3COOH, DMF, CH_3OH) and on conditions of electrooxidation the corresponding acetates, formiates or ethers of vicinal halogen-hydrins, halogenamines or, perhaps *vic*-dibromoderivatives and epoxides are obtained [48], (3-52):

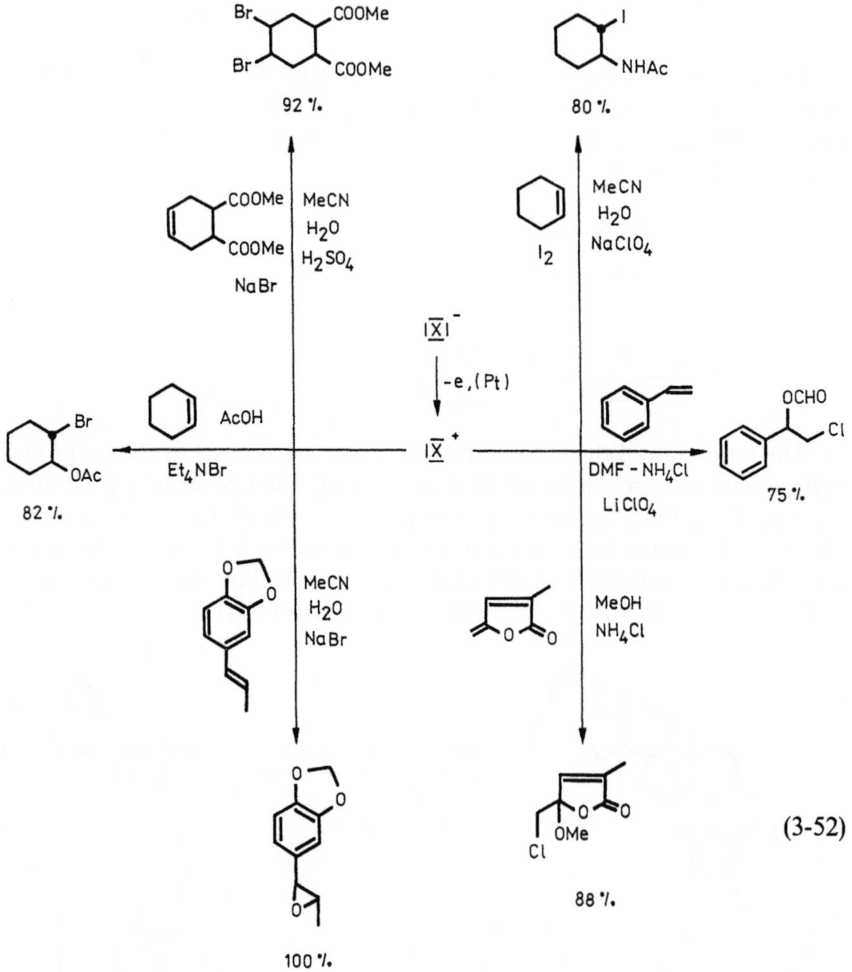

$$(3\text{-}52)$$

This reaction was made use of on industrial scale for the synthesis of halogeno-hydrins and oxirans. In the synthesis of oxirans the charge passed carries out "double work": i.e. both the hypohalo acid generated at the anode, is made use of and the alkali hydroxide generated at the electrode. The halogenide ion is continuously regenerated in the reaction mixture (3-53):

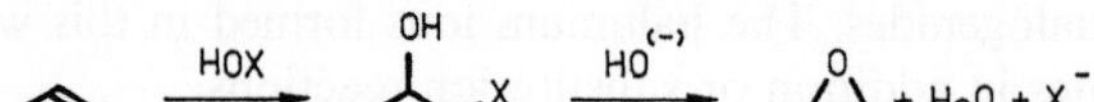

$$(3\text{-}53)$$

Electrooxidative halogenation can be applied as a simple method for a regio-selective synthesis of α-halogenoketones and their acetals from the corresponding enolethers and enolacetates [49], (3-54)

$$(3\text{-}54)$$

94 %.   95 %.

2-Bromocyclododecanone [49]

A solution of 30 mg (0.134 mmol) of (1-cyclododecenyl)acetate and 19.7 mg (0.201 mmol) of NH_4Br in acetonitrile (6 ml) and water (2 ml) was electrolyzed in an undivided cell equipped with two platinum electrodes (3 cm^2) under a constant current of 6.7 mA.cm^{-2} at 3.0–5.0 V (anode voltage 0.75 V vs Ag wire) at 20–25°C. After 2.3 F mol^{-1} of electricity was passed, the mixture was concentrated, and the residue was taken up in benzene-ethylacetate (1:1). The usual work-up gave 33.3 mg (95%) of 2-bromocyclododecanone, mp 50.5–51.5°C (pentane).

By the right choice of the halogenide ion, of its concentration in the two-phase electrolytic system and by measuring the exchanged charge one can achieve a situation in which the preceding reaction can be used in regio- and chemoselective "ene" chlorination of isoprenoids [50], (3-55):

$$(3\text{-}55)$$

91 %.

The anodically generated halonium ions are also used as electrophilic agents in order to synthesize O-, N-, C-halogenoderivatives from the corresponding substrates [48], (3-56)

$$(3\text{-}56)$$

X = CHO, Ac, CN, NO$_2$

3.1.6 Oxidation of Amines

Simple aliphatic amines have relatively low oxidation potentials and are therefore easily oxidized at an anode (E_p: $(C_3H_7)_2NH$: 1.20 V; $(C_3H_7)_3N$: 1.02 V vs NHE). The course of oxidation depends on the structure of the amine, on the anode material and on the composition of the electrolyte [51], (3-57)

$$(R-CH_2)_3N \xrightarrow{-e} (R-CH_2)_3N^{(+\bullet)} \xrightarrow{-H^+} (R-CH_2)_2N-\overset{\bullet}{C}H-R \xrightarrow{-e}$$

$$(R-CH_2)_2N-\overset{(+)}{C}H-R \rightleftharpoons (R-CH_2)_2\overset{(+)}{N}=CH-R \xrightarrow[-H^+]{H_2O}$$

$$\longrightarrow (R-CH_2)_2NH + R-CH=O \tag{3-57}$$

The electrooxidation of tertiary aliphatic amines on a carbon or on a platinum electrode and in the presence of an equivalent amount of water leads therefore to a cleavage of the C–N bond the result of which is the formation of a secondary amine and of an aldehyde (3-57). The rate at which the alkyl group is split off and yields an aldehyde, decreases with a decreasing stability of its carbanion in the sequence: $CH_3 > C_2H_5 > C_3H_7 \gg$ cyclohexyl $>$ isopropyl. The formation of benzylmethylamine as the main product in the oxidation of benzyldimethylamine also corresponds to this sequence [52], (3-58).

$$Ph-CH_2-NMe_2 \xrightarrow[(Pt),\ 1,05V(SCE)]{-2e,\ MeOH,\ Bu_4NBF_4}$$

$$\longrightarrow Ph-CH=O + Ph-CH_2NHMe + Ph-CH_2-NH_2$$
$$\qquad\quad 7\ \% \qquad\qquad\quad 71\ \% \qquad\qquad\quad 10\ \% \tag{3-58}$$

The mechanism is interpreted by an electron transfer from the electron pair of nitrogen to the anode. By the following transformations of the resulting cation radical involving deprotonation and a further one-electron oxidation one arrives at an iminium ion which by adding water is transformed to an aldehyde and a secondary amine (3-57). Depending on the amount of the passed charge the final products could be ammonia and aldehyde. By choosing the reaction conditions the oxidation of primary amines can be orientated to azo derivatives [53] and to nitriles [54], (3-59):

$$C_7H_{15}-C\equiv N \xleftarrow[Ni(OH)_2,\ -6e]{H_2O-MeCN-KOH} R-CH_2-NH_2 \xrightarrow[(Pt),\ -4e]{THF-LiClO_4} C_4H_9-N=N-C_4H_9$$

$$\qquad\qquad R=C_7H_{15} \qquad\qquad\qquad R=C_3H_7 \qquad\qquad \begin{array}{c}33\ \%\\ (3\text{-}59)\end{array}$$

The interpretation of the above formation of azo derivatives and nitriles is also based on the primary oxidation of the free electron pairs on the nitrogen atom (3-60):

$$RCH_2-N=N-CH_2R$$

$$\uparrow$$

$$RCH_2NH-NHCH_2R$$

$$\uparrow 2x$$

$$R-CH_2-\ddot{N}H_2 \xrightarrow{-e} R-CH_2-\overset{(+\cdot)}{N}H_2 \xrightarrow{-H^+} RCH_2\overset{\cdot}{N}H$$

$$\downarrow -e$$

$$R-CH=\overset{\cdot}{N} \xleftarrow{-e} R-CH=NH \xleftarrow{-H^+} RCH_2\overset{(+)}{N}H$$

$$\downarrow -e$$

$$R-CH=\overset{(+)}{N} \xrightarrow{-H^+} R-C\equiv N \tag{3-60}$$

The electrooxidation of aromatic amines depends considerably on reaction conditions. In strongly acidic media a mixture of 1,4-benzoquinone with benzidine results, in neutral media semidine is formed and in alcoholic solutions azobenzene (3-61):

$$\tag{3-61}$$

More important from the point of view of synthesis is the oxidation of mixed aliphatic-aromatic amines. The oxidation of N,N-dimethylaniline may be directed in order to prepare N-methoxymethyl-N-methylaniline or N,N-bis-(methoxymethyl)aniline. Both products demethoxylate under the influence of Lewis acids and the resulting iminium intermediate may react "in situ" with electron rich olefins giving tetrahydroquinolines [55], (3-62)

$$(3\text{-}62)$$

A considerable importance has the electrochemical methoxylation of amides and carbamates giving rise to stable α-methoxy derivatives [56, 57], (3-63) in relatively high yields, i.e. to products which are "chemically" prepared from amides or carbamates, from the aldehyde (practically only formaldehyde) and methanol (3-64). Electrochemically, however, they are prepared in higher yields even from amides which do not react in chemical processes.

$$(3\text{-}63)$$

2-Methoxy-1-methoxycarbonylpiperidine [57]

Into a 50 ml electrolysis cell fitted with two carbon electrodes were placed 0.05 mol of 1-methoxycarbonylpiperidine and 0.005 mol of tetraethylammonium p-toluene-sulfonate as electrolyte and 32 ml of methanol as a solvent. The constant current (0.5 A) was passed through the cell which was externally cooled with water. After 2 F mol^{-1} of electricity was passed, 50 ml of water was added to the reaction mixture and it was extracted with three portions of ether. The combined organic layer was dried on magnesium sulfate overnight. After removing the magnesium

sulfate by filtration, the ether was distilled off and the residue was distilled.
2-Methoxy-1-methoxy-carbonylpiperidine, bp 60°C (0.5 kPa), was obtained in
72% yield.

$$R^1-\underset{O}{\overset{\parallel}{C}}-NH-R^2 \ + \ R^3-CH=O \ + \ CH_3OH \longrightarrow$$

$$R^1-\underset{O}{\overset{\parallel}{C}}-\underset{\underset{R^3}{|}}{\overset{\overset{R^2}{|}}{N}}-CH-OCH_3 \xrightarrow[-MeOH]{H^+} R^1-\overset{O}{\overset{\parallel}{C}}-\underset{\underset{R^2}{|}}{\overset{(+)}{N}}=CH-R^3 \longrightarrow$$

$$\xrightarrow[-H^+]{H-\text{Nu}} R^1-\overset{O}{\overset{\parallel}{C}}-\underset{\underset{R^2}{|}}{N}-\underset{\overset{|}{\text{Nu}}}{CH}-R^3 \tag{3-64}$$

α-Methoxyamides and -carbamates generate iminium ions under the influence of an acid which react then with sufficiently "nucleophilic" substrates (aromates, alkenes, C-acids etc.) by an electrophilic substitution [58] which is called amidoalkylation (3-65):

$$\tag{3-65}$$

By elimination of methanol from methoxylated carbamates stable and sufficiently nucleophilic "encarbamates" can be prepared which react with electrophilic reagents. By this procedure 1-carbmethoxypiperidine was transformed to

3-pyridinecarbaldehyde [59], (3-66):

$$(3\text{-}66)$$

In an analogous way as in methoxylation, anodic acetoxylation (generally acyloxylation) of amides and carbamates can be also carried out. The course of the anodic reaction between the carboxylic acids and dimethylformamide depends on the kind of the supporting electrolyte [60]. In presence of lithium and tetraalkylammonium ions the acyloxylation of the methyl group occurs whereas in presence of triethylamine the so-called Kolbe dimers are formed (3-67):

$$(3\text{-}67)$$

The following example is only an intramolecular version of the acyloxylation reaction which takes place in the anodic oxidation of the (1-carbmethoxy-3-ethyl-3-piperidyl)acetic acid [61], (3-68):

$$(3\text{-}68)$$

α-Methoxylated amides and carbamates are also used as equivalent aldehyde reagents, e.g. in the Strecker synthesis of α-aminoacids [62], (3-69)

$$(3\text{-}69)$$

As far as the electrooxidation of carbamates is performed in presence of water, α-hydroxy derivatives are obtained which may be used directly as aldehyde equivalents, e.g. in the Wittig synthesis (3-70)

$$(3-70)$$

3.1.7 Electrooxidation of Ions

By the anodic oxidation of carboxylate anions two types of reactive intermediates can be obtained: radicals which dimerize (the so-called Kolbe reaction) and carbocations which can eliminate a proton, transform to isomeric carbocations and react with nucleophiles (the so-called Hofer-Moest reaction [63, 64], (3-71):

$$(3-71)$$

The formation of both intermediates is preceded by a single-electron oxidation of the carboxylates. The acyloxyl radical formed in this way decarboxylates and an alkyl radical is obtained which may be oxidized to the carbocation (3-72):

$$R - COO^{(-)} \xrightarrow{-e} R - COO^{\cdot} \xrightarrow{-CO_2} R^{\cdot} \xrightarrow{-e} R^{(+)} \qquad (3-72)$$

The "single-electron" and the "two-electron" mechanism of the anodic oxidation of carboxylates is affected by a number of factors: the anode material and its potential, the current density, the solvent, the supporting electrolyte, the temperature and, last but not least, by the structure of the carboxylic acid. A number of publications have been devoted to the influence of these factors; these papers represent a substantial contribution to literature in the field of organic electrochemistry.

The single-electron oxidation and – thus also the formation of the Kolbe dimer – is supported by a higher current density, higher concentration of carboxylates and by lower temperature. Among solvents methanol and a mixture of water with methanol are most frequently used. In combination with

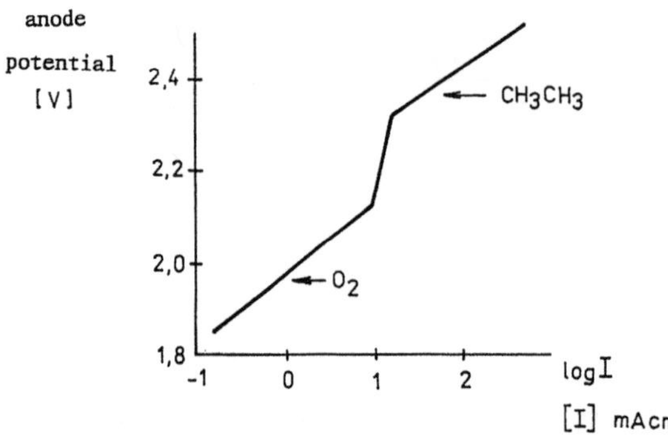

Fig. 3.2. Dependence of the potential of a platinum anode on the logarithm of current density in the oxidation of acetate ions

the supporting electrolytes the following systems are used: methanol–sodium salt of the acid; methanol–sodium methoxide; methanol–water–sodium hydroxide; methanol–triethylamine–pyridine. When using an organic solvent and water whose relative content is higher than 4% a sudden decrease occurs in the formation of the Kolbe dimer. The most suitable anodic material is platinum, anodes made from gold have been also used, as well as iridium, palladium, PbO_2 and glassy carbon. A very important parameter for a successful performance of the Kolbe synthesis is the current density on the anode. The dependence of the potential of a platinum electrode on the logarithm of current density is demonstrated in Fig. 3.2 for the oxidation of acetates in an aqueous solution. It is evident from Fig. 3.2 that at current density equal to about $10\ \mathrm{mA\ cm^{-2}}$ the reaction mechanism suddenly changes; this change appears in an abrupt increase of the anode potential. Whilst at a lower current density the electrolysis of water occurs and oxygen is evolved, at higher current densities ethane is the main product; it is formed by dimerization of methyl radicals adsorbed on the electrode surface (3-73):

$$2\ CH_3COO^- \xrightarrow[-2\ CO_2]{-2e} 2\ CH_3^{\bullet}{}_{(ads)} \longrightarrow CH_3-CH_3$$

$$(3\text{-}73)$$

The above example shows also how the kinetic effects enable the course of a reaction occurring in a direction which is thermodynamically unrealistic.

Kolbe dimerization is widely applied [65] in the synthesis of alkanes from alkanoic acids (3-74) and in the preparation of esters of alkanedioic acids from the corresponding alkyl hydrogen esters.

$$2\ CH_3-(CH_2)_n-COO \xrightarrow[MeOH]{-2e,\ -2CO_2} CH_3-(CH_2)_{2n}-CH_3$$

$$60-90\ \%$$

$$(n = 5-15)$$

$$(3\text{-}74)$$

An industrial application has been achieved in the case of production of sebacic acid diethylester from hydrogenethyladipate ($n = 4$, $R = C_2H_5$), (3-75)

$$ROOC-(CH_2)_n-COO^- \xrightarrow{-2e,\,-2\,CO_2} RO_2C-(CH_2)_{2n}-CO_2R$$

$$45-95\,\%.$$
$$(n = 4-16)$$

$$(3-75)$$

General Procedure for Electrolysis of Salts of Carboxylic Acids

The acid (0.2 mol) is electrolyzed in methanol (100 ml) containing enough sodium methoxide to neutralize 2% of the acid in a cell equipped with smooth platinum plates (4×2.5 cm) placed a few centimeters apart. The temperature is maintained at $40-50\,°C$ by external cooling and a current of $1.5-2.0\,A$ is maintained until the electrolyte becomes alkaline. This point is reached when about $20-40\%$ more current than that calculated is used. Removal of the solvent is followed by extraction with ether and purification of the product.

If aqueous methanol is used, petroleum ether is added to form a layer which dissolves the product formed. This product would otherwise coat the electrodes.

The "mixed" Kolbe dimerization was applied in the synthesis of a number of pheromones, e.g. of (Z)-11-hexadecenylacetate [66], (3-76):

$$C_4H_9-C\equiv C-(CH_2)_3-COOH \;+\; HOOC-(CH_2)_6-COOCH_3 \longrightarrow$$

$$\xrightarrow[CH_3OH,\,KOH]{-2e} C_4H_9-C\equiv C-(CH_2)_9-COOCH_3 \xrightarrow{\substack{1.\,H_2/Pd \\ 2.\,LiAlH_4 \\ 3.\,CH_3COCl}}$$

$$\longrightarrow \underset{C_4H_9}{\overset{H \qquad H}{\diagup\!\diagdown}}(CH_2)_{10}-OCOCH_3$$

$$(3-76)$$

or of the aggregation pheromone of the nun-moth (Disparlure sp. [67]), (3-77)

$$(3-77)$$

If one of the carboxylic acids in the molecule contains a double bond in a suitable position with respect to the carboxylic group an intramolecular addition of the primarily formed radical on to the double bond occurs; the newly formed radical reacts further in the sense of the mixed Kolbe dimerization [68], (3-78):

$$\text{(3-78)}$$

Interesting synthetic possibilities are also offered by procedures in which radicals are generated by anodic oxidation of carboxylate ions; first, these radicals are intermolecularly added on to the double bond of the unsaturated compound and only later the adduct radicals formed in this way mutually dimerize, or recombine with radicals formed on the anode [69], (3-79):

$$\text{(3-79)}$$

The "two-electron" oxidation of carboxylate ions is preferred on carbon anodes and is performed in a mixture of dipolar aprotic solvents with water in neutral or in alkaline media. In particular, acids having electron donating substituents on the α-carbon undergo this mechanism; the substituents stabilize the originating carbocations. The latter react then with nucleophiles present in the electrolyte and yield the products [70], (3-80) of methoxylation:

$$(3-80)$$

of acetoxylation [71], (3-81):

$$(3-81)$$

and of amidation [72], (3-82):

$$(3-82)$$

They may also undergo a rearrangement and the newly resulted carbocations are transformed to a stable molecule by a proton elimination; this reaction was made use of e.g. in the synthesis of ($\pm$)-muscone [73], (3-83):

$$(3-83)$$

The same rules as those derived for the electrooxidation of carboxylate ions hold also for the oxidation alkaneborate ions $(R_3BOY)^-$ which result from trialkyl-boranes and alkali hydroxides or alkoxides [74], $(Y = H, CH_3) - (3-84)$:

$$(3-84)$$

In this way dodecane was prepared from trihexylborane (3-85):

$$(C_6H_{13})_3 B \quad \xrightarrow[\text{KOH, MeOH}]{-2e} \quad C_6H_{13}-C_6H_{13} \qquad (3\text{-}85)$$
$$76\ \%.$$

The relatively low values of oxidation potentials of carbanions ($PhCH_2Li$: $E_p = -1.45$ V (SCE), THF, $(CH_3CO)_2CH_2$: $E_{1/2} = 0.49$ V, Fe/Fe^+, DMSO; $C_2H_5OOC-CH_2-COOC_2H_5$: 0.39 V, Fe/Fe^+, DMSO) facilitate their anodic oxidation. The radicals resulting in oxidation may dimerize or react in an aldolization reaction with unsaturated compounds.

The yields of dimers resulting by oxidation of carbanions of malonic acid [75] range between 20–55% (3-86):

$$\qquad (3\text{-}86)$$

More satisfactory yields of dimers are obtained in the anodic oxidation or organometallic compounds [76], (3-87):

$$2\ R-MgBr \quad \xrightarrow{-2e} \quad R-R \qquad (3\text{-}87)$$

$$R = C_5H_{11}\ (55\text{–}60\%);\ C_6H_5\ (55\%);\ C_{18}H_{37}\ (54\%)$$

In presence of unsaturated compounds an addition of primarily resulting radicals occurs on the double bond; by the following oxidation of the radical adducts the carbocations are formed which, in case of addition of 1,4-dicarbonyl compounds on to the ethylvinyl ether, are stabilized by cyclization and yield after following deprotonation dihydrofuran derivatives [77], (3-88):

$$\qquad (3\text{-}88)$$

$$R = Ac,\ COOMe$$

3.1.8 Oxidation of Aromatic Systems

Simple aromatic compounds (benzene, naphtalene, toluene, xylene) have relatively very positive oxidation potentials and are therefore more suitable substrates for indirect electrochemical oxidations. In the direct oxidation of an aromatic compound an electron transfer occurs from the π-electron system to the anode with the formation of a cation radical which, in general, either can dimerize or, by a reaction with nucleophiles yields substitution products (3-89):

$$(3\text{-}89)$$

Sufficiently "nucleophilic" aromatics represent the most suitable substrates for anodic dimerization which is moreover preferred by absence of nucleophiles in the electrolytic system. The electrochemically generated radical cation reacts in the sense of a electrophilic substitution with a further substrate molecule and the newly formed radical cation undergoes a further single-electron oxidation combined with deprotonation giving rise to the corresponding dimer [78], (3-90):

$$(3\text{-}90)$$

As far as the structural conditions are satisfied the anodic dimerization can also take place as an intramolecular reaction [79], (3-91). This reaction was made use of in the following synthesis:

$$(3\text{-}91)$$

A more complicated synthetic reaction of this type is the transformation of the tetrahydroisoquinoline alkaloids to alkaloids with a morphinane framework [80], (3-92):

$$(3\text{-}92)$$

The course of anodic nucleophilic substitution on the aromatic nucleus (acetoxylation, acetamidation, methoxylation etc.) is usually interpreted by the ECEC mechanism. A chemoselective substitution on the nucleus takes place only with aromatics and those derivatives of them whose substituents do not contain hydrogen atoms in the benzylic position.

The anodic acetoxylation of aromatics only exceptionally leads to the corresponding acetoxy derivatives [81] in yields exceeding 50% (3-93 and 3-94).

$$(3\text{-}93)$$

Higher yields of the oxidation products are achieved in trifluoroacetoxylations [82], (3-94).

$$(3\text{-}94)$$

This is why the trifluoroacetoxylation was also applied for an indirect synthesis of phenol from benzene the relatively high oxidation potential of which complicates a selective anodic oxidation of benzene to phenol or to hydroquinone. The phenyltrifluoroacetate resulting in the oxidation is sufficiently stable under conditions of electrooxidation and is not subjected to a follow-up oxidation. The required phenol is liberated from it by alkaline hydrolysis [83], (3-95):

$$(3\text{-}95)$$

The anodic oxidation of aromatic compounds in presence of acetonitrile yields products of acetamidation [84], (3-96).

$$\text{(3-96)}$$

85 %.

In the presence of ammonium nitrate or N_2O_4 nitroaromatics are formed [85], (3-97):

$$\text{(3-97)}$$

91 %.

In the anodic oxidation of alkylaromatics containing hydrogen atoms in the benzylic position, two competitive reactions proceed: a nucleophilic substitution on the aromatic nucleus or in the benzyl position of the alkyl group (3-98).

$$\text{(3-98)}$$

The preferential formation of the respective product can be effected by choosing suitable reaction conditions or by specially arranging the experiment. Thus, in order to prepare 2,4,6-trimethylphenylacetate, anodic acetoxylation of 1,3,5-trimethylbenzene is carried out in presence of hydrogenation catalyst by which the undesirable acetoxymethyl derivative is retransformed to the starting substrate. In this way a uniform oxidation product is obtained in a good chemical but a poor electrochemical yield [86], (3-99).

$$\text{(3-99)}$$

The anodic oxidation of the methyl group in aromatics can be directed, depending on the potential of the working electrode, on the composition of the electrolyte and on the amount of the passed charge toward the oxidation state of a benzylalcohol, of an aldehyde or of an acid (3-100):

$$\mathrm{Ar-CH_3} \xrightarrow{-2e} \mathrm{Ar-CH_2-OAc} \xrightarrow{-2e} \begin{array}{c}\mathrm{Ar-CH(OAc)_2}\\ \text{or}\\ \mathrm{Ar-CH{=}O}\end{array} \xrightarrow{-2e} \mathrm{Ar-COOH} \qquad (3\text{-}100)$$

The selective oxidation of methylaromatics to benzylacetates is achieved at lower potentials, carbon is recommended as anode material. A suitable electrolytic system for electrolytic acetoxylation is acetic acid or a solution of acetic acid and *tert*-butyl alcohol in mixture with one of the following supporting electrolytes: sodium or ammonium acetate or tetraalkylammonium perchlorate, tetrafluoroborate or tosylate. An addition of Cu^{2+} or Co^{2+} ions is also recommended. The metal ions facilitate the electron transfer from the aromatic substrate to the anode (i.e. they lower its oxidation potential) and accelerate thus the rate of formation of the intermediate benzylcarbocation [87], (3-101):

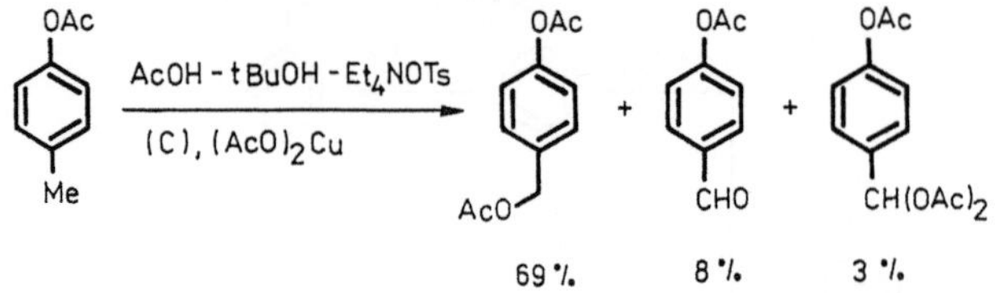

(3-101)

By the subsequent anodic oxidation of the benzylacetates aldehydes are obtained. The oxidation degree of the functional group resulting by the oxidation of the methyl group can be controlled by the potential of the working electrode as follows from the example of the electrooxidation of 2-hydroxy-3-methoxy-5-methylbenzaldehyde. In the same electrolyte at a lower potential of the anode the corresponding benzylacetate results as the main product; by a further electrolysis at a higher potential 4-hydroxy-5-methoxy-1,3-benzenedicarbaldehyde is obtained [88], (3-102):

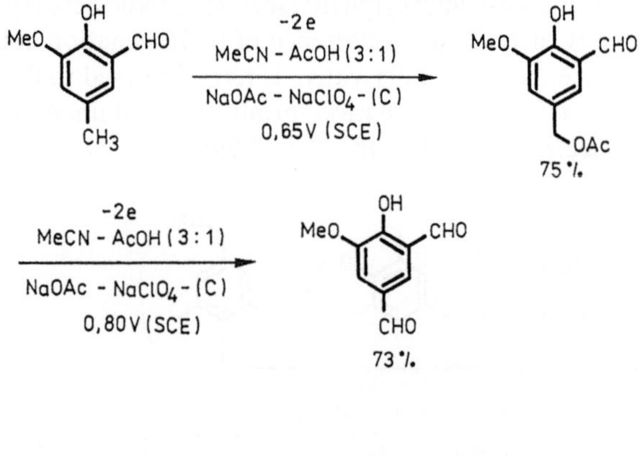

(3-102)

The anodic oxidation of benzylalkylethers proceeds very easily, in particular in those which contain a methoxy group or, in general, an electron-donating group at position 2 or 4 of the benzene nucleus. These groups lower the oxidation potential of benzylethers by 0.4–0.5 V [89], (3-103):

$$MeO-\text{C}_6\text{H}_4-CH_2OMe \xrightarrow[-2e]{MeOH-CH_2Cl_2-H_2SO_4} MeO-\text{C}_6\text{H}_4-CH{=}O \qquad (3\text{-}103)$$

97 %.

This fact is also made use of in synthesis for the electrochemical setting free of the alcoholic group whose hydroxyl was protected in the form of corresponding benzylether [90], (3-104)

$$\xrightarrow[-2e,\ LiClO_4]{MeCN-H_2O\ (Pt)} \qquad (3\text{-}104)$$

89 %.
+
$$MeO-\text{C}_6\text{H}_4-CH{=}O$$

The easy electrooxidation of the benzyl positions makes possible the modification of the carbon framework in steroids, e.g. the introduction of a double bond [91], (3-105):

$$\xrightarrow[\substack{-2e,\ MeOH,\\2,6\text{-lutidine}}]{CH_2Cl_2,\ NaClO_4}$$

100 %.

$$(3\text{-}105)$$

The electrooxidation of p-substituted toluenes followed by acid-catalyzed hydrolyses of the intermediate acetals affords the corresponding aromatic aldehydes in good yields [92], (106):

$$R-\text{C}_6\text{H}_4-CH_3 \xrightarrow{-4e}$$

1. MeOH, AcOH, NaBF$_4^-$
2. H$_3$O$^+$ → AcO-C$_6$H$_4$-CH=O   72 %.

1. MeOH, KF
2. H$_3$O$^+$ → 80 - 85 %.

$$(3\text{-}106)$$

4-Acetoxybenzaldehyde [92]

Into a 100 ml undivided electrolytic cell equipped with a thermometer and carbon rod electrodes was placed 80 ml of acetic acid – methanol (1:9) containing 0.05 mol of sodium tetrafluoroborate as a supporting electrolyte and 0.02 mol of (4-methylfenyl)-acetate. Stirred with a magnetic bar and cooled with a water bath, the solution was electrochemically oxidized at the constant current of 200 mA (current density: 1.6 A dm⁻²). After 8.0 F mol⁻¹ of electricity was passed through the reaction system, the reaction mixture was poured into 100 ml of an aqueous sodium chloride solution and extracted with three 50 ml portions of ether. The combined ethereal solution was evaporated to remove the solvent and then poured into 100 ml of 10% aqueous sulfuric acid. The mixture was stirred for 3 hr at room temperature, and extracted with three 50 ml portions of ether. The combined solution was washed with 10% aqueous sodium bicarbonate solution and dried over anhydrous magnesium sulfate. After filtration of the solvent, the residual oil was subjected to distillation to give the corresponding 4-acetoxybenzaldehyde, bp 115–116°C (0.66 kPa), in 72% yield.

In the anodic oxidation of methylaromatics performed in acetonitrile and in presence of perchlorates ($(C_2H_5)_4N^+$, Li^+, Na^+) the corresponding acetamido derivatives are obtained [93]. The oxidation is carried out on platinum and, in particular, on carbon anodes, the relative content of water in the electrolyte should not exceed 1% (3-107):

$$(3\text{-}107)$$

1,4-Disubstituted benzene derivatives with electron donating substituents easily undergo anodic oxidation in which quinol structures are formed [94], (3-108):

$$(3\text{-}108)$$

Useful intermediates in organic synthesis are bisacetals of 1,4-benzoquinone and of its derivatives prepared by the oxidative addition of methoxyls to the corresponding derivatives of 1,4-dimethoxybenzene [95, 96], (3-109):

(3-109)

The facile oxidizability of hydroquinone ethers and esters is made use of in the electrochemical generation of acyl cations and in the preparation of esters. The reaction is based on the anodic oxidation of 4-hydroxyphenylalkanoates in the solution of the corresponding alcohol. The transient oxidation product is decomposed to p-benzoquinone and to the ester [97], (3-110):

(3-110)

The anodic cyanation of aromatics gives only exceptionally good results as regards the corresponding nitriles [98], (3-111):

(3-111)

Very interesting, however, is the course of the anodic cyanation of 1,4-dimethoxybenzene in which an ipso substitution of the methoxyl group by the cyanide ion occurs and 4-methoxybenzonitrile is obtained in a high yield [99], (3-112)

(3-112)

A substitution of the aromatic nucleus by the electrochemically generated halonium ions only occurs if the oxidation potential of the halogenide is lower than the potential of the aromatic substrate. The other way round, first the anodic oxidation of the aromatic substrate occurs, followed by the reaction of the resulting carbocations with halogenides of the supporting electrolyte [100], (3-113):

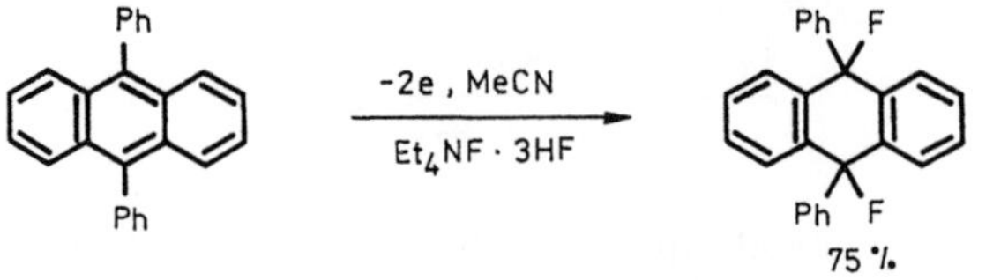

$$\text{(3-113)}$$

Among the anodic oxidations of heterocyclic compounds closest attention has been paid to methoxylation, acetoxylation and hydroxylation of furan and of its derivatives. The methoxylation in which 2,5-dihydro-2,5-dimethoxyfurans result which are cyclic acetals of 1,4-dicarbonyl compounds, were studied in detail; for this reason they are widely applied in the synthesis of various types of compounds [101], (3-114)

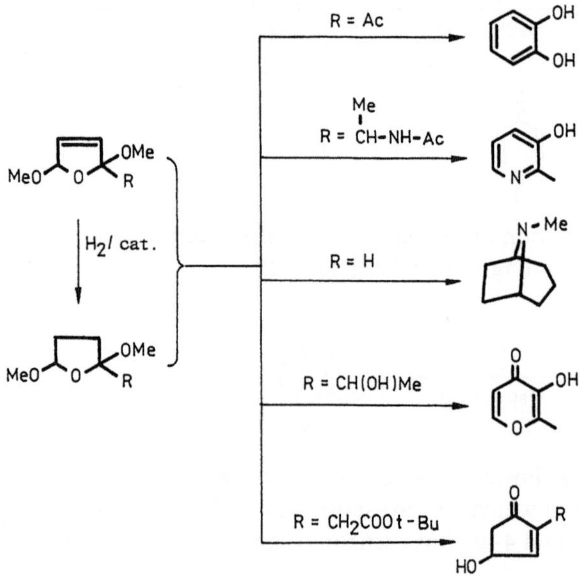

$$\text{(3-114)}$$

The methoxylation is performed by electrolysis of the furan substrate in a methanol solution and its reaction path depends on the structure of the starting compound and on the supporting electrolyte. The very frequently used ammonium bromide is suitable only for methoxylation of those furan derivatives which carry electron donating substituents; a primary oxidation of the bromide ion and the addition of bromine to the diene system are assumed. The reaction is one of the first examples of indirect anodic oxidation (cf. Sect. 3.3). The methoxylation of 2-acetyl- and of 2-carbalkoxyfurans must be performed in the presence of sulfuric acid as supporting electrolyte in order to obtain acceptable yields of dihydrodimethoxyfurans; the oxidation starts by an electron transfer from the furan nucleus to the anode (an ECEC mechanism). In an oxidation in an alkaline media (CH_3OH–KOH) a participation of methoxyl radicals in the reaction is also possible (3-115).

$$(3\text{-}115)$$

The selectivity of product formation is also affected by the amount of the passed charge and by the current density. A selective transformation of furfuryl alcohol to its dimethoxy-dihydro derivative is achieved at current densities smaller than $0.05\ \mathrm{A\,cm^{-2}}$ and a charge of $2\ \mathrm{F\,mol^{-1}}$. At higher current densities $(0.20\ \mathrm{A\,cm^{-2}})$ and charge magnitudes of 4–$6\ \mathrm{F\,mol^{-1}}$ the cycle is opened, the substituent at position 2 is split off and a methylester of the dimethylacetal of 3-formylpropenic acid results [102], (3-116)

$$(3\text{-}116)$$

(2,5-Dimethoxy2,5-dihydro-2-furyl)methanol [102]

1.00 g of 2-Furylmethanol was dissolved in 20 ml of methanol containing 0.10 g of tetraethylammonium perchlorate as a supporting electrolyte. The solution was electrolyzed at 15–16 °C under a current density of 0.033 A cm⁻² in a compartment cell equipped with two platinum foil electrodes (2 × 3 cm²). After 2.1 F mol⁻¹ of electricity was passed, the reaction mixture was concentrated under reduced pressure, taken in ethylacetate, washed with aqueous $NaHCO_3$ and brine, and dried. After evaporation of the solvent the residue was distilled to give (2,5-dimethoxy-2,5-dihydro-2-furyl)methanol, bp 82–85 °C (1.0 kPa), in 91% yield.

Methyl-(Z)-4,4-dimethoxy-2-butenoate [102]

A solution of 3.00 g of 2-furylmethanol and 0.10 g of tetraethylammonium perchlorate in 20 ml of methanol was electrolyzed under the same conditions as in the preceding case. After 8.0 F mol^{-1} of electricity was passed, the reaction mixture was worked up in the usual way and methyl-(Z)-4,4-dimethoxy-2-butenoate, bp 74–78 °C (1.6 kPa), was obtained in 75% yield.

The same product may be also obtained in the oxidation of furan-2-carboxylic acid; by changing the supporting electrolyte it is possible to prepare also the opposite stereoisomer [103], (3-117).

$$(3\text{-}117)$$

Whereas the methoxylation only occurs in the positions of 2 and 5 of the furan nucleus, the acetoxylation of 2,5-dimethylfuran occurs on the methyl groups [104], (3-118)

$$(3\text{-}118)$$

The anodic oxidation of 3,4-disubstituted furans in acetonitrile in the presence of water leads to the formation of 2,5-dihydro-2,5-dihydroxyfurans [105], (3-119)

$$(3\text{-}119)$$

The degree of oxidation of methoxylation products of pyrrole depends on the character of substituents located on the nitrogen atom. *N*-methylpyrrole undergoes a 4-electron oxidation whereas 1-carbmethoxypyrrole yields only the product of a 2-electron oxidation [106], (3-120):

$$(3\text{-}120)$$

Of great promise are the electrochemical oxidations of 5-membered heterocycles (pyrrole, thiophene, furan) carried out in inert atmosphere in acetonitrile solutions and in presence of tetraalkylammonium salts. During the oxidation electrically conductive polymers are formed on the surface of platinum electrodes; their analysis demonstrated that a single anion belongs to four monomeric units of the heterocycle [107, 108], (3-121):

$$(3\text{-}121)$$

A polypyrrole film prepared by the potentiostatic oxidation of pyrrole in an acetonitrile solution of $(C_2H_5)_4NBF_4$ at 0.8 V (vs SCE) exhibits semiconducting properties; in the oxidized (doped) state it is conductive, dark brown till black. The charge of the anion of the supporting electrolyte (BF_4^-) compensates for the charge of the oxidized, onium form of polypyrrole; this corresponds to a single electron per 4 monomeric units. In the cathodic potential range polypyrrole in its reduced (undoped) form is nonconducting, its colour is yellow. The oxidation potentials of monomeric heterocycles and the electric conductivity of the polymeric films obtained are demonstrated in Table 3.2.

In the case of polypyrrole films their conductivity and many other properties may be modified by suitable substitution on nitrogen (by an alkyl or by a phenyl) on the one hand and by the choice of the anion which passes into the doped form from the supporting electrolyte.

Table 3.2. Oxidation potentials of monomers and the electric conductivity of polymeric films

Compound	Oxidation potential (V vs SCE)	Conductivity (S cm^{-1})
pyrrole	+ 0.8	30–100
indole	+ 0.8	5×10^{-3}–10^{-2}
thiophene	+ 0.9	10–100
furan	+ 1.85	10–80

3.2 Direct Cathodic Reductions

3.2.1 Reductions of Functional Groups

Electrochemical hydrogenations of multiple C–C bonds can proceed indirectly, electrocatalytically, or by a direct electron transfer on to the multiple bond. In the electrocatalytic method hydrogen is developed in acid media on cathodes activated by Pt or Pd or in alkaline media on Raney nickel cathodes. Hydrogen then catalytically hydrogenates the double bond. In the other method a direct reduction of the double bond occurs in an electroorganic process (E) a radical anion is primarily formed which in a chemical process (C) yields a radical by a protonation; by a further E process this yields an anion which in a follow-up C process leads to the corresponding dihydrogen derivative (3-122):

$$(3\text{-}122)$$

The different mechanism of both methods of electrochemical hydrogenation is also reflected in stereochemistry of the resulting products. Whilst in the electrocatalytic hydrogenation prevalently the *cis*-dihydro derivatives result, in the direct reduction occurring by an ECEC mechanism the *trans*-dihydro derivatives are formed as the main product [109, 110], (3-123):

$$(3\text{-}123)$$

The electrochemical reduction of benzene and of its derivatives carried out in primary amines (ethylenediamine, methylamine) as solvents and in presence of lithium chloride yields products which also result by "chemical" reduction under the influence of alkali metals (Na, Li) in solutions of primary amines (Birch reduction). Benzene proper can be electrochemically reduced under the above conditions to the 1,4-cyclohexadiene in an undivided cell or to cyclohexene in a divided cell (3-124):

$$\text{(3-124)}$$

The different course of the reaction is caused by the isomerization of the primarily formed 1,4-cyclohexadiene to 1,3-cyclohexadiene which, as a conjugated diene is electroactive again and undergoes a further cathodic reduction. The above mentioned isomerization proceeds in a divided cell under the influence of lithium methylamide, $LiNHCH_3$, which is produced at the cathode. In an undivided cell the above amide is decomposed by reaction with methylamine hydrochloride resulting at the anode; this is why the main reaction product is 1,4-cyclohexadiene [111].

The different reducibility of an isolated double bond enables a chemoselective electroreduction of the aromatic nucleus in alkenylaromatics with a non-conjugated double bond [112], (3-125):

$$\text{(3-125)}$$

On the other hand, in case of conjugation, e.g. with cinnamic acid, a selective reduction of the double bond (3-126) occurs at a mercury cathode:

$$\text{(3-126)}$$

Dialkylacetylenes, however, are electrochemically active and by the reduction in a methylamine solution of lithium chloride they yield *trans*-alkenes whereas by electroreduction in an ethanolic solution of H_2SO_4 they give rise to *cis*-alkenes [113, 114], (3-127):

$$\text{(3-127)}$$

On industrial scale the electroreduction of multiple bonds was applied e.g. for the production of dihydrophtalic acid from phtalanhydride (3-128):

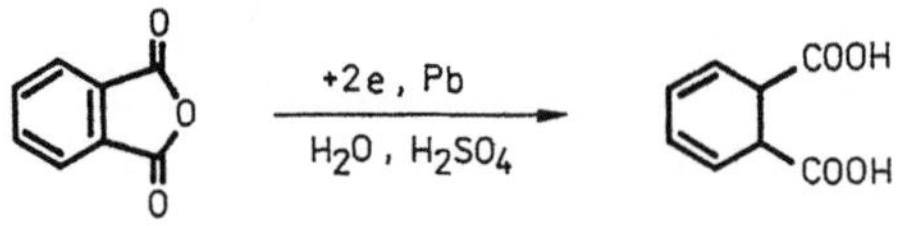

$$\text{(phthalic anhydride)} \xrightarrow[\text{H}_2\text{O , H}_2\text{SO}_4]{+2\,\text{e , Pb}} \text{(benzene-1,2-dicarboxylic acid, COOH, COOH)} \qquad (3\text{-}128)$$

or of piperidine by the reduction of pyridine [115], (3-129):

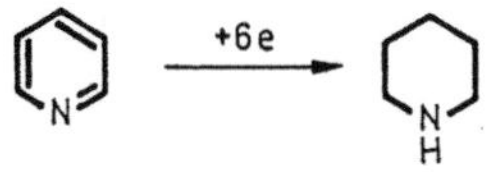

$$\text{(pyridine)} \xrightarrow{+6\,\text{e}} \text{(piperidine, N-H)} \qquad (3\text{-}129)$$

The electrochemical reduction of organic halogenoderivatives was studied both from the point of view of mechanism and from that of the synthetic use. The reaction products are based on the participation of radical anions, radicals and carbanions which all in dependence on the conditions of electrolysis (composition of the supporting electrolyte, the cathode potential, presence of compounds with multiple bond) can undergo a number of reactions (protonation, recombination, addition, reaction with the cathode).

The reductive cleavage of the C–Hal bond is irreversible. The ease of reduction decreases in the order $R\text{–}I > R\text{–}Br > R\text{–}Cl \gg \gg R\text{–}F$, further *tert*-$R\text{–}X > sec\text{-}R\text{–}X > prim\text{-}R\text{–}X$ and finally $R\text{–}CX_3 > R\text{–}CHX_2 > R\text{–}CH_2X$.

An important role in the electroreduction of halogeno-derivatives is also played by the electrode material and by the potential at which the reduction is carried out. The final products of the reduction of halogenoalkanes at a mercury electrode are mostly alkanes. Thus in the electroreduction of 1-iododecane in dimethylformamide at a cathode potential -1.6 V (vs SCE) didecylmercury is obtained whilst at -2.21 V (vs SCE) decane (3-130) results. On the other hand, 1-bromodecane yields only decane in the reduction [116].

$$C_{10}H_{22} \xleftarrow[-2,14\,\text{V}]{+2\,\text{e , (Hg) , DMF}} C_{10}H_{21}-I \xrightarrow[-1,61\,\text{V}]{+\text{e , (Hg) , DMF}}$$

$$\longrightarrow C_{10}H_{21}-Hg-C_{10}H_{21} \qquad (3\text{-}130)$$

The appearance of didecylmercury points to the primary formation of a radical anion or of a radical with a sufficient stability which makes a reaction with mercury possible. Only at more negative potentials a second electron is rapidly accepted.

The above mentioned sequences in the reactivity of the individual types of halogenoderivatives or the dependences of their reduction potentials on the structure can be made use of in the selective reductions of the individual halogens in presence of the others [117], (3-131)

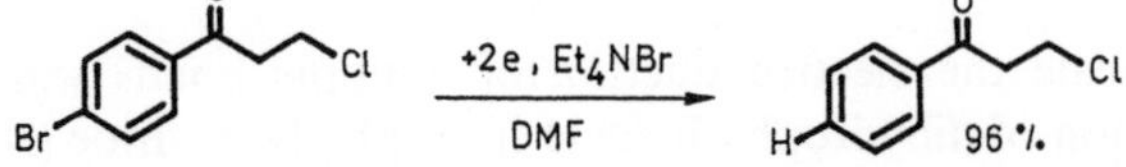

$$\xrightarrow[\text{DMF}]{+2\,\text{e , Et}_4\text{NBr}} \qquad 96\ \% \qquad (3\text{-}131)$$

or to a partial "electrohydrogenolysis" of the higher halogenated substrates [118], (3-132):

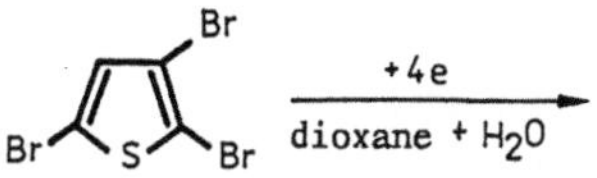

$$\text{(structure)} \xrightarrow[\text{dioxane} + \text{H}_2\text{O}]{+4e} \text{(structure)} \tag{3-132}$$

The selective reduction of halogenoderivatives is achieved even without a potentiostat if a salt is chosen as the supporting electrolyte which also plays simultaneously the role of a "potential buffer". In this way one can e.g. achieve the reduction of a $-CCl_3$ group to $-CHCl_2$ or up to CH_2Cl. In the first case the easily reducible ammonium nitrate is chosen as the supporting electrolyte, in the second case $(CH_3)_4NCl$ is applied which is reduced with more difficulty [119], (3-133)

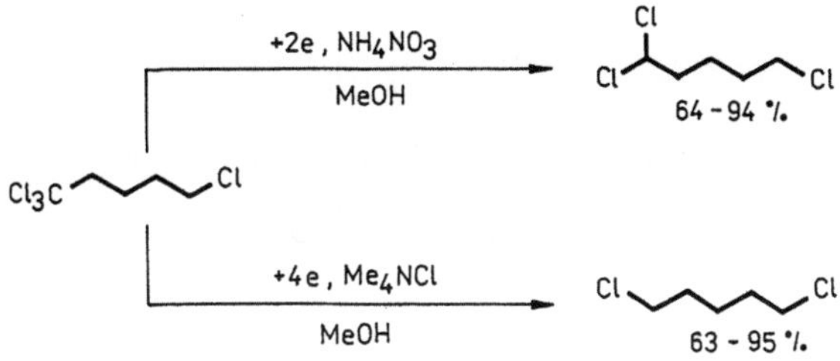

$$\tag{3-133}$$

The steric accessibility of the individual halides is also reflected in the reduction selectivity [120], (3-134)

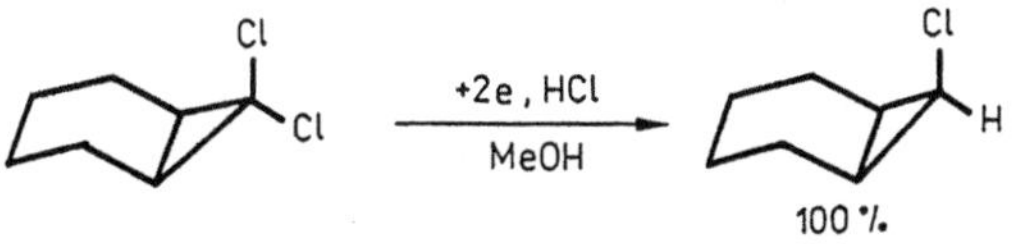

$$\tag{3-134}$$

In a number of cases the selectivity of halide reduction is affected not only by the electrode material but also by its "surface modification". Thus the highly selective hydrogenolysis of tetrachloropicolinic acid to 3,6-dichloropicolinic acid – a plant growth stimulator – is achieved on a carbon cathode on which spongy silver has been deposited. Other electrodes, even silver cathodes treated in another way are not active [121], (3-135):

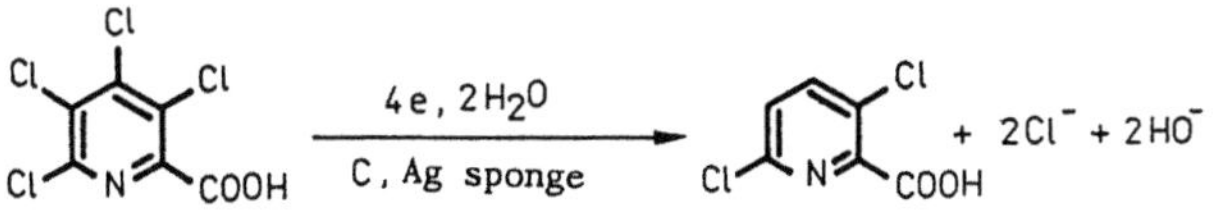

$$\tag{3-135}$$

Cathodic reductions of halogeno derivatives by which purposefully organometallic compounds are prepared are performed on the so-called sacrificed electrodes from the corresponding metal and separated by a diaphragm from the

Pt anode [122], (3-136)

$$Hg(CH_2CH_2CN)_2 + 2HI$$

$$\uparrow (Hg), 2H^+, +2e$$
$$(H_2SO_4)$$

$$Sb(CH_2CH_2CN)_3 + 3HI \xleftarrow[K_2HPO_4]{(Sb), +3e, 3H^+} \boxed{I-(CH_2)_2-CN} \xrightarrow[(H_2SO_4)]{(Pb), +4e, 4H^+} Pb(CH_2CH_2CN)_4 + 4HI$$

$$\downarrow (Sn), +6e$$
$$NaOH$$

$$(NCCH_2CH_2)_3Sn-Sn(CH_2CH_2CN)_3 \qquad (3\text{-}136)$$

In a similar manner to the C–Hal bonds the strongly polar bonds in ammonium salts also undergo electroreduction; the ammonium salts are transformed to tertiary amines [123], (3-137) and (3-138):

$$(3\text{-}137)$$

88 – 90 %.   n = 1–3

$$(3\text{-}138)$$

Phosphonium salts yield tertiary phosphines [124], (3-139):

$$(3\text{-}139)$$

72 %.

A nitro group, particularly one that is bonded to an aromatic nucleus, belongs to the groups whose reduction on a cathode is extremely easy. Because of its low reduction potential it can be selectively reduced even in presence of other functional groups, e.g. of a carboxylic group. A high selectivity is achieved when working with controlled potential, but also by the choice of current density or by the choice of the cathode material, this being usually a metal with a low hydrogen overvoltage. The course of the reduction also depends on the pH of the solution and the products can be amines, 4-aminophenols, azoxy, azo and hydrazo derivatives, as well as benzidines [111].

Nitrobenzene can be selectively reduced in a divided cell (Ni cathode and Pb anode) to aniline or hydrazobenzene (3-140):

$$
\begin{array}{l}
\text{(Ni), NaOH} \\
\underline{\qquad\qquad} \quad \phi-N=N-\phi \;+\; 4\,H_2O \\
8e,\,8H^+
\end{array}
$$

$$
\phi-NO_2 \quad \xrightarrow[\;6\%\,HCl\,,\;PbCl_2\;]{\;(Ni),\,6H^+,\,6e\;} \quad \phi-NH_2 \;+\; 4\,H_2O \tag{3-140}
$$

$$
\begin{array}{l}
\text{(Ni), 10H}^+,\,10\,e \\
\underline{\qquad\qquad} \quad \phi-NH-NH-\phi \;+\; 4\,H_2O \\
\text{AcONa, EtOH}
\end{array}
$$

The reduction of aromatic nitro compounds to azoxy, azo and hydrazo aromatics is performed in alkaline media at cathodes with a low hydrogen overvoltage. Since nitro $(-0.54\,V)$, azoxy $(-0.55\,V)$ and azobenzene $(-0.31\,V$ vs SCE) have very near reduction potentials the reduction usually leads to the poorly soluble azoxy derivative which is deposited during electrolysis (3-141):

$$
Ar-NO_2 \;\longrightarrow\; Ar-\underset{\underset{O}{|}}{N}=N-Ar \;\longrightarrow\; Ar-N=N-Ar \;\longrightarrow\;
$$

$$
\longrightarrow\; Ar-NH-NH-Ar \;\longrightarrow\; H_2N-Ar-Ar-NH_2 \tag{3-141}
$$

The further reduction of azoxy derivatives is carried out with an excess of alcohol or in the presence of tetraalkylammonium salts of aromatic sulfonic acids (McKee's salts) in order to increase the solubility of the azoxy derivatives. As a rule, the azo compound is obtained, accompanied by the hydrazo derivative which can be by an oxidation in open air transformed to the azo species. By a further reduction of the azo compound at lower current density hydrazo derivatives result which in strongly acidic electrolytes rearrange to benzidines.

The reduction of o- and p-nitrophenols and nitroanilines occurs also in alkaline media because the temporarily formed nitroso derivative is transformed to a quinoid form which is very easily reduced to the corresponding aminophenol or phenylenediamine (3-142).

$$(3\text{-}142)$$

Nitrobenzene reduced at a cathode with a low hydrogen overvoltage (Pt, carbon) in a strong mineral acid yields phenylhydroxylamine which is transformed to 4-aminophenol.

By electrolysis at controlled potential and in a diluted acid, in order to prevent the transformation, the corresponding phenylhydroxylamine can be obtained which in case of a suitable substitution of the starting substrate intramolecularly condenses with a carbonyl group; the reaction has been made use of e.g. in the synthesis of quinoline-*N*-oxide (3-143)

$$(3\text{-}143)$$

High yields of amines are obtained in the reduction of nitroaromatics carrying a group in the para position which prevents a rearrangement [125], (3-144).

$$(3\text{-}144)$$

By a suitable choice of conditions (cathode potential, temperature, pH of the solution) in aromatic polynitro compounds one can carry out a selective reduction of the individual nitro groups up to the oxidation degrees NHOH or NH_2 and obtain so the corresponding phenylhydroxylamines, nitroanilines, aromatic polyamines or, perhaps, the aminophenols [126], (3-145).

$$(3\text{-}145)$$

These possibilities may be seen on polarographic curves of such compounds with two electroactive groups which are reduced in two waves with different half-wave potentials.

Amines or hydroxylamines result also by reduction of aliphatic nitrocompounds [111]. In general, one can say that the oxidation degree of amines is achieved by electrolysis at a higher temperature (70 °C), higher current density and at cathodes with sufficiently high hydrogen overvoltage (Hg, Pb); (3-146) and (3-147).

$$(3\text{-}146)$$

$$(3\text{-}147)$$

From the point of view of organic synthesis the selective reduction of nitro-alkanes to the corresponding hydroxylamines is interesting and necessary; it is performed at lower temperatures or at controlled potentials of the mercury cathode (3-148),

$$(3\text{-}148)$$

or the reduction is performed at a cathode with a low hydrogen overvoltage (Pt, Ni, graphite), (3-149)

$$\text{C}_6\text{H}_{11}\text{-NO}_2 + 4e + 4H^+ \xrightarrow[\text{(Ni)}]{\text{MeOH}-\text{HCl}} \text{C}_6\text{H}_{11}\text{-NHOH} + H_2O \qquad (3\text{-}149)$$

94,4 %.

Primary amines [127] can be obtained conveniently by electrochemical reduction of aldoximes (3-150) and ketoximes (3-151)

$$\xrightarrow[\text{+4e, NaOH}]{\text{(Pb), K}_2\text{HPO}_4-\text{H}_2\text{O}} \qquad (3\text{-}150)$$

57 %.

$$\xrightarrow[\text{+4e}]{\text{(Pb), KOH}-\text{H}_2\text{O}} \qquad (3\text{-}151)$$

85 %.

or by the reduction of nitriles; from the adipic acid dinitrile, 1,6-hexanediamine (3-152) can be prepared in an undivided cell,

$$NC\text{-}\cdots\text{-}CN \xrightarrow[\text{+8e, H}_2\text{O}]{\text{(steel), NaOH}-\text{CuSO}_4} H_2N\text{-}\cdots\text{-}NH_2 \qquad (3\text{-}152)$$

83 %.

acrylonitrile can be transformed by electroreduction in a cell with a diaphragma to allylamine (3-153).

$$\text{CH}_2\text{=CH-CN} \xrightarrow[\text{4e}]{\text{(Pb), 30 \% H}_3\text{PO}_4} \text{CH}_2\text{=CH-CH}_2\text{-NH}_2 \qquad (3\text{-}153)$$

In the reduction of aromatic and heteroaromatic nitriles at a mercury cathode the mechanism and hence also the products are strongly affected by the pH of the solution; at constant pH also by the potential of the electrode [128], (3-154).

pH < 7

$$\xrightarrow[\text{(Hg)}]{4e, 4H^+}$$

pH ≫ 7

$$\xrightarrow[]{2e, H_2O} \quad + CN^{(-)} + HO^{(-)}$$

$$(3\text{-}154)$$

The primary step is the uptake of the first electron. At a pH of around 7 this reaction is followed by two competitive reactions: uptake of further electrons and protonation is faster and through the sequence ECEC leads to the product RCH_2NH_2. If the potential is shifted to more negative values the rate of protonation does not increase whilst the rate of the electron uptake grows exponentially. The result is the splitting off of CN^- as in alkaline media.

In the reduction of benzonitriles carrying electronegative $-$ M substituents e.g. $COOCH_3$, CN which lower the electron density on the reaction centre (electron-withdrawing groups) the same mechanisms make themselves active and the same types of products result.

By electroreduction of *N*-nitroso and *N*-nitrocompounds the corresponding hydrazine derivatives are formed [127], (3-155) and (3-156):

$$(3\text{-}155)$$

$$(3\text{-}156)$$

$$X = O,\ 20\%\ H_2SO_4;\ X = NH,\ 30\%\ (NH_4)_2SO_4$$

Aldehydes and ketones are reduced to the corresponding alcohols, as a rule in alkaline and buffered media, at electrodes made of mercury, lead, zinc, or cadmium. The aldehydes are more reactive than the ketones (3-157).

$$(3\text{-}157)$$

The electrochemical method makes possible a chemoselective reaction of the carbonyl group in the presence of other functional groups [127], e.g. nitrile, ethoxycarbonyl etc. (3-158).

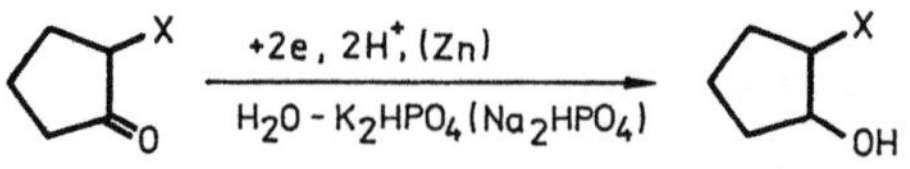

$$(3\text{-}158)$$

$$X = H\ (76.4\%);\ X = CN,\ COOEt\ (75\text{--}79\%)$$

The mechanism of the reduction of alkylarylketones in partly aqueous solutions in a pronounced way depends on pH of the solution. Whereas in acid media a two-electron reduction is preferred which leads to secondary alcohols, in alkaline media a single electron reduction proceeds resulting in the formation of *vic*-diols [129], (3-159):

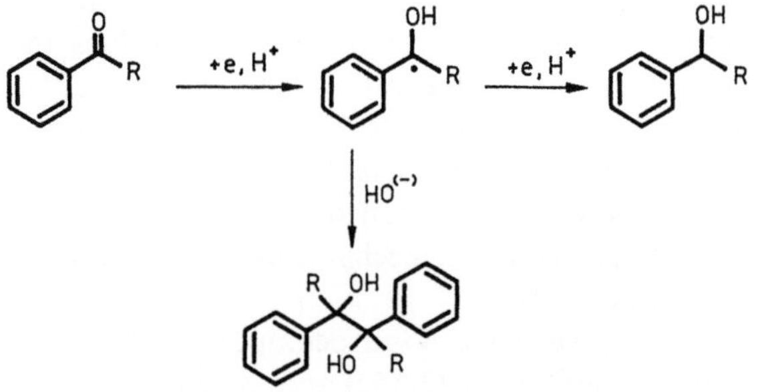

$$(3\text{-}159)$$

If D-(−)-ephedrine hydrochloride is used as supporting electrolyte an enantioselective course of acetophenone reduction has been observed, but the resulting (*R*)-(+)-2-phenylethanol is obtained in a low optical yield (4.2%), (3-160)

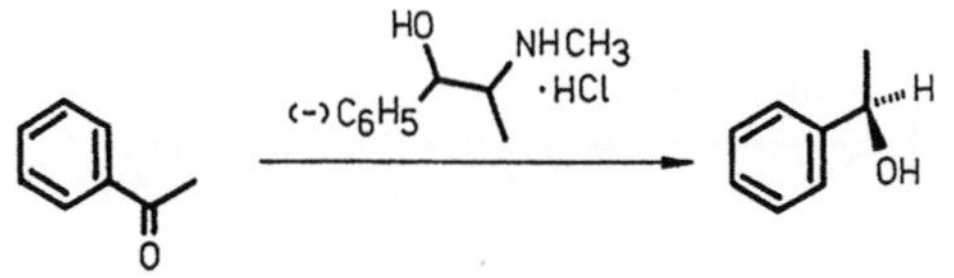

$$(3\text{-}160)$$

A substantially higher enantioselectivity in the formation of one enantiomer (75%) has been achieved in the reduction of 2-acetylpyridine on a mercury cathode on which strychnine was adsorbed [121], (3-161).

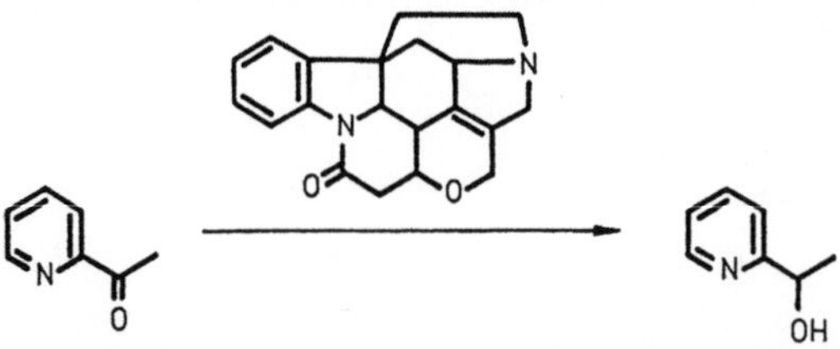

$$(3\text{-}161)$$

From the economic point of view a pairing of the cathodic reduction of glucose to sorbitol with the oxidation of the former to gluconic acid taking place at the anode is advantageous [130]; in this arrangement "a single charge performs double work" (3-162):

$$(3-162)$$

The cathodic reduction of 4-oxo-2,2,6,6-tetramethylpiperidyl oxide is interesting because its course can be controlled by the choice of the cathode material and by the amount of the charge exchanged [127], (3-163).

$$(3-163)$$

The electrochemical reduction of carboxylic acids and of their esters to aldehydes and alcohols only proceeds with those structures in which the carboxylic group is activated by another carboxylic group or by an aromatic or heteroaromatic nucleus. This means that aliphatic acids are not reduced at a cathode. In industry, such a reaction has been used for the production of glyoxylic acid from oxalic acid (3-164).

$$(3-164)$$

In a similar way to that in aliphatic-aromatic ketones the aromatic nucleus is not attacked in the electroreduction of aromatic acid esters [131], (3-165).

$$(3-165)$$

The oxidation degree of the resulting products can be affected by the cathode material, by the pH value of the catholyte and by the exchanged charge [127], (3-166).

$$(3\text{-}166)$$

Among the other acid derivatives the electrochemical reduction of amides of aliphatic and aromatic acids as well as that of lactams should be mentioned. All these reactions give rise to the corresponding amines [132], (3-167).

$$(3\text{-}167)$$

35 %.

N-Ethyl-1-methyl-2-(3-trifluoromethylphenyl)ethylamine [133]

A solution of 1-(3-trifluoromethylphenyl)-2-propanone (2.02 g; 10 mmol) in an electrolyte consisting of 70% aqueous ethylamine (10 ml, 150 mmol), 2N HCl (37.5 ml) and ethanol (50 ml) is purged with N_2 and electrolyzed at 10°C with controlled potential of -1.75 V (vs SCE) in a divided cell with mercury pool as working electrode and a platinum anode. The current drops from an initial 50 mA to 2–3 mA when the necessary 1960 A. s have been consumed. Then the electrolyte is acidified by adding 2M HCl (to pH 1). After the ethanol has been removed under reduced pressure, the aqueous solution is extracted with ether (2 × 50 ml) and NaOH is added (to pH 14). The alkaline solution is extracted with ether (3 × 50 ml), and the combined organic phase is dried ($MgSO_4$). Removal of the solvent under reduced pressure gives 2.0 g (87%) of N-ethyl-1-methyl-2-(3-trifluoromethylphenyl)ethylamine.

Substituted amines also result in the electroreduction of azomethines (Schiff's bases); it is not necessary to prepare them in advance; it is sufficient to reduce electrolytically a mixture of an aldehyde with an amine: the transiently resulting amine is reduced to the corresponding amine [133], (3-168).

$$(3\text{-}168)$$

90 %.

3.2.2 Reactions of Cathodically Generated Species

3.2.2.1 Additions

Carbanions or radicals resulting by cathodic reduction of suitable substrates, in particular halogenoderivatives, are used for functionalizing a number of organic substrates by adding the carbanions on to unsaturated compounds, especially to those with activated double bonds.

By reduction of tetrachloromethane in dimethylformamide the trichloromethyl carbanion results which, by reacting with aldehydes or ketones, or in an addition to an activated double bond in acrylonitrile forms the corresponding trichloromethyl derivatives [134, 135]. In the presence of chloroform the following addition takes place:

$$CCl_4 + 2e \longrightarrow {}^{(-)}CCl_3 + Cl^{(-)}$$

$$\tag{3-169}$$

It can be looked upon as a chain reaction and the current yields of the products reach values up to $10^3\%$ (3-169).

The reaction was made use of, for example, in the construction of sugar molecules; by adding $CCl_3^{(-)}$ to D-glyceraldehyde acetonide a mixture of epimers of the corresponding trichloromethanol is obtained which are in the following step transformed by electrohydrogenolysis to the tetrose oxidation step. Depending on the following treatment of the epimeric dichloromethylalcohols one can arrive to the D-erythrose or D-threose isopropylidene dimethylacetal or to D-erythrulose isopropylidene acetate [136], (3-170).

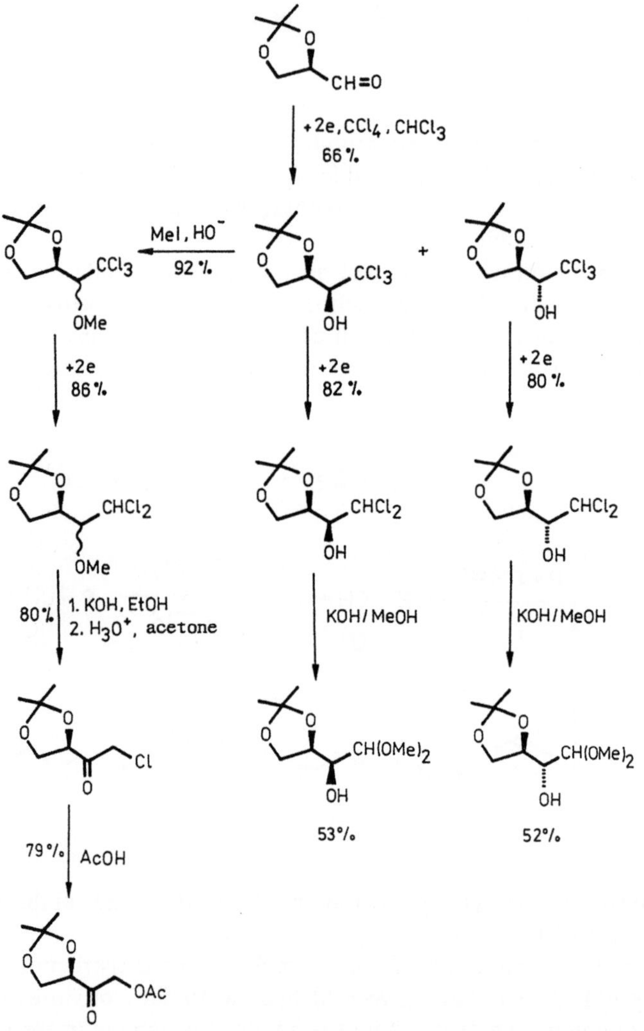

$$(3\text{-}170)$$

In a way analogous to the addition of tetrachloromethane, the "chain" addition of a mixture of ethyltrichloroacetate and ethyldichloroacetate to carbonyl compounds (3-171) is carried out

$$R-CH=O \; + \; CCl_3-COOC_2H_5 \; \xrightarrow[CHCl_2COOC_2H_5]{+2e}$$

$$\longrightarrow \; R-\underset{\underset{OH}{|}}{CH}-CCl_2-COOC_2H_5 \qquad (3\text{-}171)$$

or the addition of other geminal trichloromethyl derivatives [137]. The alcoholate anions resulting in their addition need not necessarily be transformed to
the final product by protonation only. In aprotic solvents intramolecular
nucleophilic substitution may occur under the formation of cyclic ethers [138],
(3-172).

$$(3-172)$$

40 %

The alcoholate ions resulting from the addition of carbanions generated by
cathodic reduction of β-halogenoesters or of the corresponding β-trimethylam-
monium derivatives, also cyclize intramolecularly to the corresponding lactones
(3-173).

$$(3-173)$$

$$X = I, MeEt_2N^+; \quad R^1 = H, Me; \quad R^2 = C_5H_{11}, C_7H_{15}$$

As the electrophilic component for the addition of cathodically generated
carbanions carbon dioxide may also be applied; the reduction products of such
a "electrocarboxylation" reaction are carboxylic acids or their methyl esters
[139], (3-174).

$$(3-174)$$

60 %

Among the other types of ions the addition of amidic ions generated by cathodic
reduction of N-chlorocarbamates, e.g. to acrylic acid esters (3-175) may have a
synthetic significance [140].

(3-175)

The cathodic reduction of aldehydes and ketones in acid media gives rise to radicals which upon addition to unsaturated compounds yield adduct radicals which undergo a further reduction at the cathode; then by protonation give the corresponding adducts – secondary or tertiary alcohols [141], (3-176).

(3-176)

The above addition is often denoted as the mixed or the crossed hydrodimerization. The electroreductive hydrodimerization of aldehydes or ketones with ethyl acrylate is carried out in an aqueous – methanolic solution at a potential -1.20 to -1.30 V vs SCE on a mercury cathode; the primary product of the addition, the γ-hydroxyester, cyclizes to a derivative of γ-butyrolactone (3-176). Good yields of the addition products are also obtained with a carbon cathode and a buffered base solution (K_2HPO_4 etc.)

The additions carried out to a nonactivated double bond in an analogous way give relatively low yields of tertiary alcohols with the only exception of allylalcohol (3-177).

(3-177)

If, however, the crossed hydrodimerization can occur in an intramolecular way, cycloalkanols are obtained with a high degree of regio- and stereoselectivity [142], (3-178).

(3-178)

$$R^1 = CH_3-C_5H_{11}; \quad R^2 = H, (CH_3)_2CH$$

Electroreduction of Nonconjugated Olefinic Ketones [142]

In a 100 ml undivided electrolysis cell equipped with carbon rod electrodes and a reference electrode was placed a solution of 0.01 mol of nonconjugated olefinic ketone and 30 g (0.10 mol) of tetraethylammonium p-toluenesulfonate in 50 ml of methanoldioxane (1:9). Stirred with a magnetic bar and cooled with running water, the solution was electrochemically reduced at the constant current of 200 mA. After almost complete consumption of the starting ketone (about 10 F mol^{-1} of electricity was passed), the reaction mixture was poured into 200 ml of saturated solution of sodium chloride and extracted with three 100 ml portions of ether. The combined ethereal solution was dried over anhydrous magnesium sulfate and evaporated. All products were isolated by preparative GC.

3.2.2.2 Substitutions (acylation, alkylation)

The anionic intermediates resulting from the cathodic reduction of various types of organic substrates can react with acylating and alkylating agents. The acylating agents are usually anhydrides and halogenides of acids, their nitriles and dimethylformamide. In the acylation of heteroanions acid derivatives are formed. Thus, by cathodic reduction of nitroalkanes in the presence of acetanhydride, the corresponding N-alkyl derivatives of the mixed anhydride of acetic and acetohydroxamic acid were obtained [143], (3-179).

$$R-NO_2 \xrightarrow[\text{(MeCO)}_2\text{O}]{+2e,\ \text{MeCN}} R-N \tag{3-179}$$

R = Me (54%), Ph (55%)

Under analogous conditions, the disulfides yield thioacetic acid esters [144], (3-180).

$$\tag{3-180}$$

By the cathodic reduction of pyrazine 1,4-diacetyl-1,4-dihydropyrazine was obtained [145], (3-181).

$$\tag{3-181}$$

Ketones are obtained by the acylation of cathodically generated carbanions. The carbanions can be formed by the reduction of halogenoalkanes [146],

(3-182)

$$(3\text{-}182)$$

69 %.

or of compounds with an activated double bond [147], (3-183).

$$(3\text{-}183)$$

82 %.

In the reduction of 2-phenylpropene acetonitrile plays simultaneously both role of the solvent and that of the acylating agent [148], (3-184).

$$(3\text{-}184)$$

68 %.

A similar situation can be observed in the "double" electroreductive formylation of styrene which is carried out in a dimethylformamide solution [148], (3-185).

$$(3\text{-}185)$$

82 %.

For the alkylating of the electrochemically generated carbanions usually alkylhalogenides, -mesylates and -tosylates are applied. Relatively good yields of the products are obtained, particularly if the alkylation can proceed as an intramolecular process. It is interesting to compare the different regioselectivity in the intramolecular cyclization of dimethyl-4-bromobutylidenemalonate achieved by electroreductive generation of the carbanion and by the chemical generation caused by the influence of a complex hydride. In the former case dimethylcyclobutylmalonate results, in the latter case dimethylcyclopentanedicarboxylate [149], (3-186).

$$(3\text{-}186)$$

65 %.

An almost quantitative yield of (4,4,1) propellan-3-one was achieved in the electroreductive intramolecular alkylation (cyclization) of the corresponding α,β-unsaturated bicyclic ketone [150], (3-187).

$$\text{(3-187)}$$

An intramolecular alkylation of anions resulting in the electroreduction of azomethinium ions was used in the synthesis of alkaloids [151], (3-188).

$$\text{(3-188)}$$

Both types of alkylations – the inter- and intramolecular one – occur if the reduction of benzanil or azobenzene is carried out in presence of 1,4-dibromo-butane [152], (3-189).

$$\text{(3-189)}$$

3.2.2.3 Pinacolizations and Hydrodimerizations

By electroreduction of aliphatic and aromatic aldehydes and ketones at catho-des with a different hydrogen overvoltage and within a broad range of poten-tials, vicinal diols [153] (pinacols) are formed. The new C–C bond is a result of recombination of radicals or of radical ions formed by a "single electron" reduction of the carbonyl group. The aromatic carbonyl compounds give as a rule higher yields of pinacols than the aliphatic ones (3-190).

$$2\,R-CO-R' \;+\; 2e \;\xrightarrow[\;(2\,H_2O)\;]{\;2H^+\;}\; \underset{\underset{OH}{\overset{|}{\;}}}{\overset{R}{\overset{|}{C}}}\!\!-\!\!\underset{\underset{OH}{\overset{|}{\;}}}{\overset{R}{\overset{|}{C}}}$$

(3-190)

The pinacolization is very often applied in synthesis, e.g. the most suitable method for the preparation of 1,2-cyclopropanediols is an intramolecular pinacolization of 1,3-diketones [154], (3-191).

(3-191)

Relatively good yields (50%) of pinacol prepared from retinol can be achieved by an electrochemical procedure alone if diethylmalonate was applied as the source of protons [155], (3-192).

(3-192)

The azomethines [156] also dimerize under conditions of ketone pinacolization (3-193).

$$R-CH=N-R' \;\xrightarrow[\;EtOH-AcOMe-H_2O\;]{\;+2e\;}\; \underset{\underset{NHR'}{\overset{|}{\;}}}{R-CH}\!\!-\!\!\underset{\underset{NHR'}{\overset{|}{\;}}}{CH-R} \quad 88\,\%$$

$$R = 4\text{-}Cl\text{-}C_6H_4, \; R' = \text{cyclohexyl}$$

(3-193)

The intermolecular electroreductive hydrodimerization of α,β-unsaturated compounds (aldehydes, ketones, esters and nitriles) is one of the most powerful electroorganic synthetic reactions [141]. The best results were achieved in the case of commercially realized acrylonitrile dimerization to adiponitrile [157], which at the beginning was performed in divided cells, at lead cathodes and in the presence of tetraethylammonium tosylate or ethylsulfate as supporting electrolyte. The above salts ensure high conductance, prevent the reduction of acrylonitrile to propionitrile and increase its solubility in the solution to be electrolyzed. The formation of the C–C bond is interpreted by the recombination of resulting radical anions or by their conjugated addition on to the starting substrate (3-194).

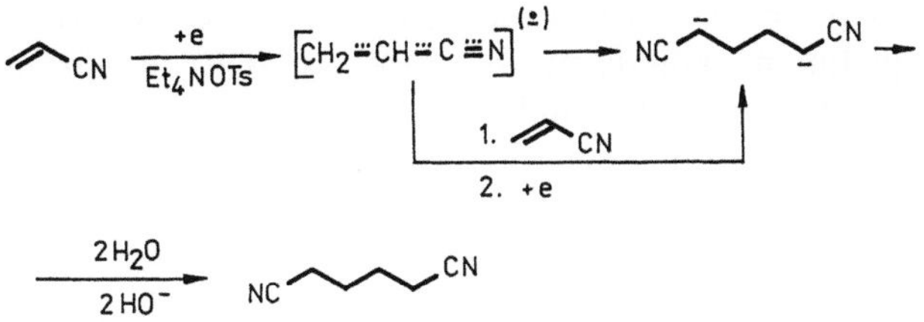

$$\text{CH}_2\text{=CH-CN} \xrightarrow[\text{Et}_4\text{NOTs}]{+e} \left[\text{CH}_2\text{=CH=C=N}\right]^{(\pm)} \longrightarrow \text{NC-CH}_2\text{CH}_2\text{CH}_2\text{-CN}^{-} \longrightarrow$$

$$\xrightarrow[\text{2 HO}^-]{2\,\text{H}_2\text{O}} \quad \text{NC-CH}_2\text{CH}_2\text{CH}_2\text{CH}_2\text{-CN} \tag{3-194}$$

Recently, the economy of the electrochemical production of adiponitrile has been improved by the technique in which the hydrodimerization is performed in undivided cells. This has been made possible by applying highly selective cadmium cathodes and an optimal composition of the electrolyte; the supporting electrolyte is hexamethylene bis(dibutylethylammonium)phosphate jointly with corrosion inhibitors (Na_2HPO_4, $Na_2B_4O_7$, Na_4-EDTA) (3-195):

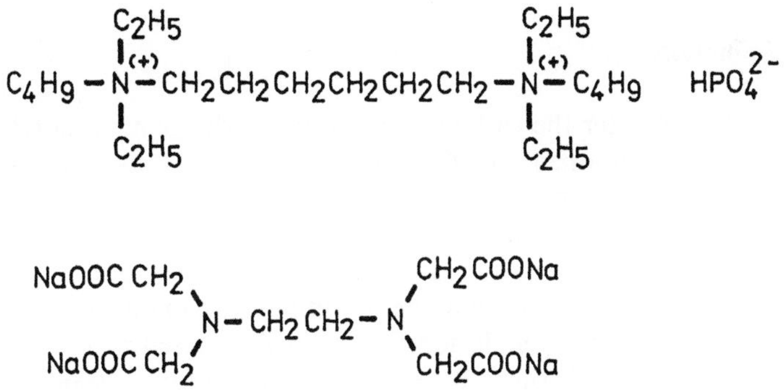

$$\tag{3-195}$$

Symmetrical α,β-unsaturated carbonyl compounds which can, after the uptake of a single electron, undergo an "intramolecular hydrodimerization" usually give high yields of the corresponding cyclic products [158], (3-196).

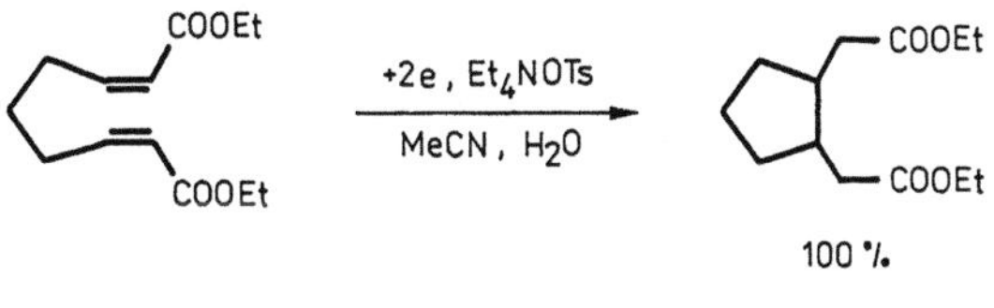

$$\xrightarrow[\text{MeCN, H}_2\text{O}]{+2e\,,\ \text{Et}_4\text{NOTs}} \tag{3-196}$$

100 %.

This reaction was preparatively used in the formation of the perhydrophenanthrene framework [159], (3-197).

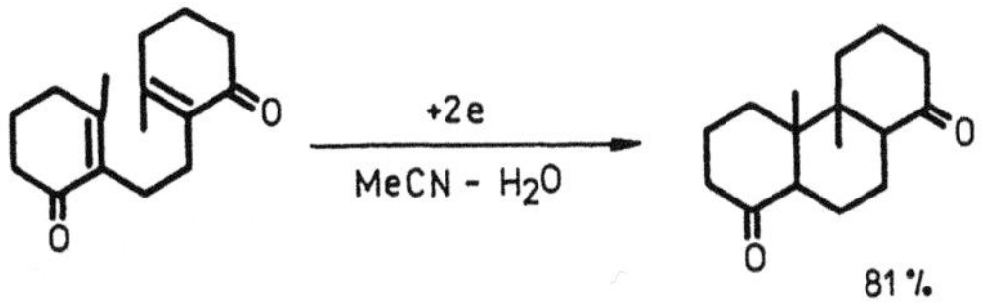

$$\xrightarrow[\text{MeCN - H}_2\text{O}]{+2e} \tag{3-197}$$

81 %.

The double bond in the fluorinated alkenes has, in a similar way as in α,β-unsaturated nitriles, an electrophilic character and easily undergoes a reduction at the cathode. In the case of e.g. octafluorocyclopentene the potentiostatic reduction at a platinum cathode at -1.80 V (vs SCE) in aprotic media (acetonitrile, dimethylformamide) and with tetraalkylammonium salts as supporting electrolyte led to the preparation of a glossy polymer which in the doped state is blue-black and exhibits the properties of a semiconductor (conductivity 10^{-3} S cm^{-1}). It is assumed to have a structure in which one tetraalkylammonium cation compensates for three monomeric units of cyclo-alkene [160], (3-198).

$$(3\text{-}198)$$

3.2.2.4 Eliminations

In the preceding chapter the cathodic reductions of the geminal polyhalogeno derivatives were mentioned in which the carbanions thus formed yielded addition products with compounds containing an activated double bond in the molecule or gave products of hydrogenolysis when proton donors were present in the reaction mixture. Of course in the absence of both types of compounds (electrophiles and proton donors) the resulting carbanion is stabilized by splitting off a further halogenide ion which, in the presence of a sufficiently "nucleophilic" alkene adds on to the double bond and forms gem-dihalogeno-cyclopropanes. From tetrachloromethane a dichlorocarbene was prepared in this way by whose addition to 2-methyl-2-butene 1,1-dichlorotrimethylcyclo-propane is formed [161], (3-199).

$$(3\text{-}199)$$

In an analogous way, difluorocarbene results from the cathodic reduction of dibromodifluoromethane; on addition it yields a *gem*-difluorocyclopropane derivative [162], (3-200).

$$(3\text{-}200)$$

As far as the structure of the substrate and the reaction conditions make it possible, the carbene thus formed may undergo an intramolecular "insertion" reaction [163], (3-201).

$$(3\text{-}201)$$

The reduction of *N,N*-dichloro-*p*-toluene sulfonamide is an example of a cathodically formed nitrene which reacts with 1,4-dioxane in an insertion reaction [164], (3-202).

$$(3\text{-}202)$$

The carbanions which result by an electrochemical reduction of *vic*-dihalogenoderivatives undergo a stereospecific *trans*-elimination of the vicinal halogen under the formation of an alkene [165, 166], in a similar manner as when using "chemical reductants" (1,2-elimination), (3-203), (3-204).

$$(3\text{-}203)$$

$$CF_2Cl-CHCl_2 \xrightarrow[\text{HOAc}]{+2e,\ NaOAc} CF_2=CHCl \qquad (3\text{-}204)$$

$$80\ \%$$

A preferred use has been found for the electrochemical dehalogenation in the synthesis of alkenes with a strong internal stress; these are usually isolated as stable cycloaddition products of a reaction with dienes [167], (3-205).

$$(3\text{-}205)$$

The electrochemical technique can be also used for the 1,2-elimination of substituents of different type: thus e.g. the elimination of a hydroxy and of an arylsulfinyl group was applied in the transformation of a methoxycarbonyl group to a vinyl group [168], (3-206)

$$R-COOMe \; + \; Ar-SO_2-CH_2MgI \; \longrightarrow \; R-CO-CH_2-SO_2-Ar \; \longrightarrow$$

$$\xrightarrow{NaBH_4} \; R-\underset{\underset{OH}{|}}{C}H-\underset{\underset{SO_2Ar}{|}}{C}H_2 \; \xrightarrow[DMF]{+2e} \; R-CH=CH_2 \qquad (3\text{-}206)$$

$$R = C_{17}H_{35} \; (82\,\%)$$

and the elimination of hydroxy and phenylmercapto groups in the homologation of carbonyl compounds [169], (3-207).

$$\xrightarrow[DMF]{+2e} \qquad 92\,\% \qquad\qquad (3\text{-}207)$$

Formally, a cathodic reduction of 1,ω-dihalogenoalkanes can be looked upon as an elimination if such a reaction proceeds by an intramolecular S_N reaction and enables the synthesis of cyclic compounds [170], (3-208).

$$\xrightarrow[THF]{+2e\,,\;Bu_4NClO_4} \qquad\qquad (3\text{-}208)$$

This method is specially convenient for the synthesis of compounds with a large internal stress – cyclopropanes and spirocyclic compounds [171], (3-209)

$$\xrightarrow[DMF,\;47\%]{+2e\,,\;-1.6V\,(SCE)} \qquad \xrightarrow[DMF,\;40\%]{+2e\,,\;-2.3V\,(SCE)} \qquad\qquad (3\text{-}209)$$

and propellanes [172], (3-210).

$$\xrightarrow[DMF]{+2e\,,\;Et_4NBr} \qquad 12\,\% \qquad\qquad (3\text{-}210)$$

For reaching the same synthetic aims one can also use esters of 1,3-diols with methanesulfonic acid which are easily procurable from 1,3-carbonyl derivatives [173], (3-211).

$$\text{(3-211)}$$

84 %

Electroreductive Synthesis of Cyclopropanes [173]

Electrochemical reduction of dimethanesulfonates was carried out in a divided cell equipped with a lead cathode and a platinum anode. To a stirred solution of dry DMF (50 ml) containing tetraethylammonium p-toluenesulfonate (5.0 g, 17 mmol) was added dropwise a solution of dimethanesulfonate (5 mmol) in dry DMF (5 ml) over the period in which 3 F mol^{-1} of electricity had been passed. The current was constant (0.2 A), and the current density was 5.7 mA cm^{-2}. The cathode potential was about -2.5 V vs SCE. During the reaction, the solution was stirred and cooled in an ice-water bath. After 4–10 F mol^{-1} of electricity was passed, the cathodic solution was poured into cold 5% hydrochloric acid (100 ml), and the product was extracted with pentane (3 × 50 ml). The combined organic layer was dried over anhydrous magnesium sulfate. After the drying agent was removed by filtration, the solvent was evaporated, and the residue oil was purified by column chromatography on silica gel (pentane) to afford cyclopropane derivatives.

3.2.2.5 Removal of Protecting Groups [174]

The electrochemical method for the removal of the protecting groups has a parallel in the above-mentioned cathodic reactions – i.e. in the electrohydrogenolysis and in the 1,2-elimination of halogenoderivatives. An advantage of this method in comparison to chemical procedures is its high selectivity and mild reaction conditions. A high selectivity is achieved by combination of suitable electrophores-protecting groups which sufficiently differ in reduction potentials ($\Delta E_{1/2}$ 200 mV). The difference in the values of reduction potentials of a benzenesulfonyl and 4-methoxybenzoyl group enables their selective cleavage and hence also their application in the protection of the respective amino group, e.g. in the molecule of L-lysine [175], (3-212).

$$(3\text{-}212)$$

In the chemoselective 1,3-etherification of glycerol, 9-phenyl-9,10-dihydro-10-oxo-9-anthracenyl (Tro)- and the 4-cyanobenzyl (3-213) were applied as protecting groups [176].

$$(3\text{-}213)$$

The above examples represent a category of protecting groups which in fact are removed by electrohydrogenolysis of a single bond. The second category of protecting groups comprises the 2,2,2-trichloroethyl group, applicable for protecting hydroxyl, amino and mercapto groups. The principle of their removal is based on the cathodic 1,2-elimination of *vic*-derivatives [177], (3-214).

$$R-O-\underset{O}{\overset{\|}{C}}-O-CH_2-CCl_3 \xrightarrow[-1,5V\,(SCE)]{+2e\,,\ MeOH} \left[R-O-\underset{O}{\overset{\|}{C}}-O-CH_2-\overset{\overline{\ }}{C}Cl_2 \quad Cl^-\right]$$

$$\longrightarrow R-O^{(-)} + CO_2 + CH_2{=}CCl_2 + Cl^- \qquad (3\text{-}214)$$

$$R = benzyl \quad (70\%)$$

Selective Removal of Protecting Groups Using Controlled Potential Electrolysis [174]

The electrolysis was carried out in a cylindrical vessel fitted with two side arms which are separated from the main compartment with coarse grade glass frits. A mercury pool electrode in the main compartment serves as the cathode (working electrode) while a saturated calomel reference electrode is placed in one side arm and a platinum sheet in the other side arm (anode, counter electrode). The electrolyte solution (0.1 M lithium perchlorate in methanol) is added to the cell and the side arms. To the solution in the main compartment (ca. 20 ml) is added 2,2,2-trichloroethyl benzoate (0.50 g, 1.97 mmol) and the mixture is degassed with argon for 15–30 min. Then the potentiostat is applied to ensure a potential difference of − 1.65 V between the mercury pool and the reference electrode. The current, initially 145 mA, drops off slowly over 2 h, reaching a steady value of ca. 3 mA. Water cooling is necessary to prevent local heating effects. The reaction mixture is separated from the mercury, concentrated at reduced pressure, and after protonation, the usual isolation via acidification, aqueous extraction, and one recrystallization, 0.21 g (87%) of benzoic acid is obtained, mp 120–121°C.

The difference between the reduction potentials of the C–Cl and C–Br bonds is applied for the selective protection of phosphoric acid esters in the synthesis of nucleotides in such a way that the 2,2,2-trichloroethyl and 2,2,2-tribromoethyl groups are applied, both groups simultaneously [178], (3-215).

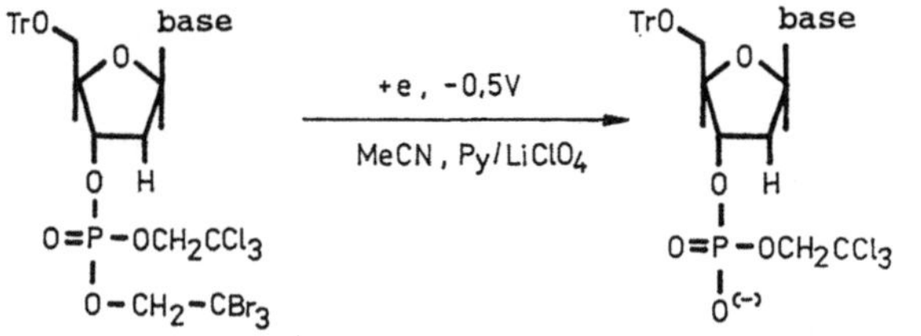

$$+e, \ -0.5V, \quad MeCN, \ Py/LiClO_4 \tag{3-215}$$

The different substitution on carbon atoms of *vic*-dibromoderivatives is reflected in the different values of reduction potentials. This fact enables e.g. the synthesis of 1,2-dibromo-1-(3-cyclohexenyl)ethane from 4-vinylcyclohexene which adds one equivalent of bromine on to a single double bond in the cycle. For this reason the initial addition of two equivalents is performed which is followed by a selective cathodic elimination of *vic*-bromines from the cyclohexane ring [179], (3-216).

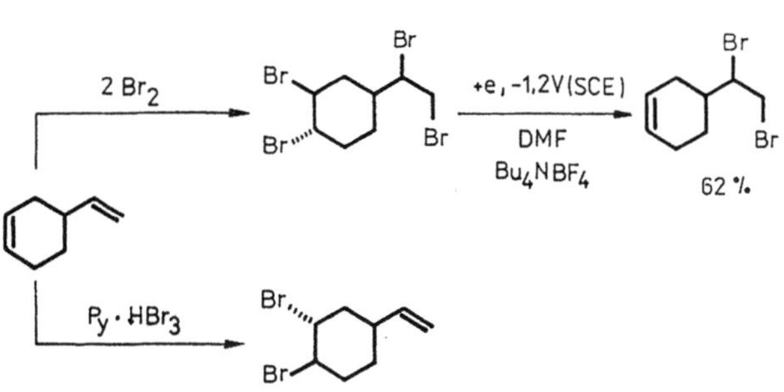

$$\tag{3-216}$$

3.3 Indirect Anodic Oxidations [180, 181, 182]

These reactions are carried out by means of the so-called mediators – i.e. electrochemically regenerable redox systems. The mediators may be fixed to the anode surface or may be a part of the mixture to be electrolyzed (of the electrolyte).

In the former case of the oxidation agent (M_{ox}) is generated on the electrode surface and fixed on it; after the reaction with the substrate (S) M_{ox} forms the product from it and is transformed to its reduced form (M_{red}) which is again continuously reoxidized to M_{ox} (cf. Fig. 3.3).

The most frequent kind of such an electrode is the so-called nickel–hydroxide electrode which is prepared by deposition of nickel(III) oxide–hydroxide on the surface of a platinum or nickel anode by electrolysing an alkaline solution of nickel(II) salts (3-217).

$$2\,[3\,Ni(OH)_2 \cdot 2H_2O] + 6\,HO^{(-)} + 3/2\,NaOH \rightleftharpoons$$

$$\rightleftharpoons 2\,[3\,Ni(O)OH] \cdot 7H_2O \cdot 3/4\,NaOH + 3\,H_2O + 6e$$

$$(3\text{-}217)$$

This electrode is especially suitable for the oxidation of primary alcohols to alkanoic acid and of primary amines to nitriles. The oxidations at Ni(O)OH electrodes are probably indirect, heterogeneous anodic dehydrogenations. The generally accepted mechanism assumes a homolytic cleavage of the α-hydrogen atom in the alcohol molecule which is adsorbed on the anode surface; under the influence of Ni(O)OH an α-hydroxyalkyl radical results in this way which is then further oxidized – directly or indirectly – to the carboxylic acid (3-218).

$$HO^- + Ni(OH)_2 \xrightarrow{\text{a fast reaction}} NiOOH + H_2O + e$$

$$RCH_2{-}OH_{dissolv.} \longrightarrow RCH_2{-}OH_{ads.}$$

$$RCH_2OH_{ads} + NiOOH \xrightarrow{\text{rate determining step}} R{-}\overset{\bullet}{C}H{-}OH + Ni(OH)_2$$

$$R{-}\overset{\bullet}{C}H{-}OH + H_2O \longrightarrow RCOOH + 3e + 3H^+$$

direct oxidation

$$R{-}\overset{\bullet}{C}H{-}OH + 3\,NiOOH + H_2O \longrightarrow RCOOH + 3\,Ni(OH)_2$$

indirect oxidation

$$(3\text{-}218)$$

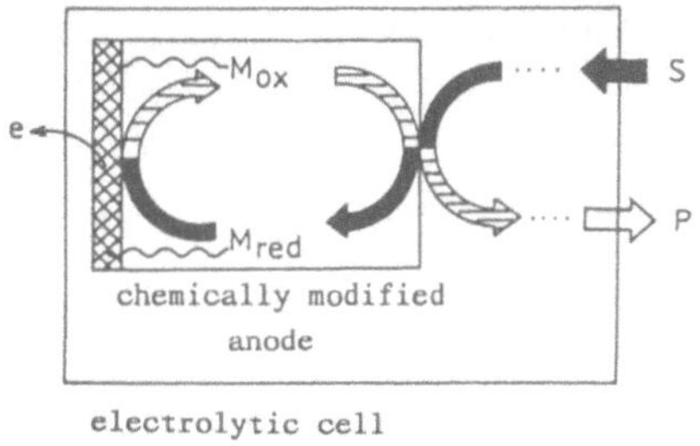

Fig. 3.3. Principle of the indirect electrochemical synthesis with use of a chemically modified anode

The oxidation of alcohols is carried out in an aqueous solution of sodium hydroxide or in potassium hydroxide in a mixture of water with *tert*-butyl alcohol. The anode is kept in the "catalytic state" by adding trace amounts of nickel(II) salts. High yield of alkanoic acids are obtained from lower alcohols at laboratory temperature, from octanol and higher alcohols not until temperatures from 60 to 70 °C are reached (3-219).

$$R{-}CH_2{-}OH \xrightarrow{Ni(O)OH,\ -4e} R{-}COOH \qquad (3\text{-}219)$$

$R = C_4H_9$	H_2O-*t*-BuOH–KOH,	25 °C	92%
C_6H_{13}	H_2O-*t*-BuOH–KOH,		91%
$C_{10}H_{21}$	H_2O–NaOH,	25 °C	27%
	H_2O–NaOH,	70 °C	87%

Oxidations at a nickel-oxide anode resemble both in selectivity and in the yields the oxidations carried out with nickel peroxide. They successfully compete with oxidations by permanganate or potassium bichromate as regards the simplicity of the application the ease of processing the reaction mixture, and the yields. In their technological application, the troubles with the liquidation of the reduction products may be omitted, e.g. that of manganese(IV) oxide and of chromium salts. For this reason the procedure has been successfully applied in the "wasteless" technology for one step in the production of vitamin C, in the oxidation of 2,3:4,6-di-*O*-isopropylidene-L-sorbose to 2,3:4,6-di-*O*-isopropylidene-L-gulonic acid (3-220). The oxidation can be carried out in undivided cells without controlling the anode potential; the good conductivity of the electrolyte makes it possible to reach high currents at a low voltage across the electrodes.

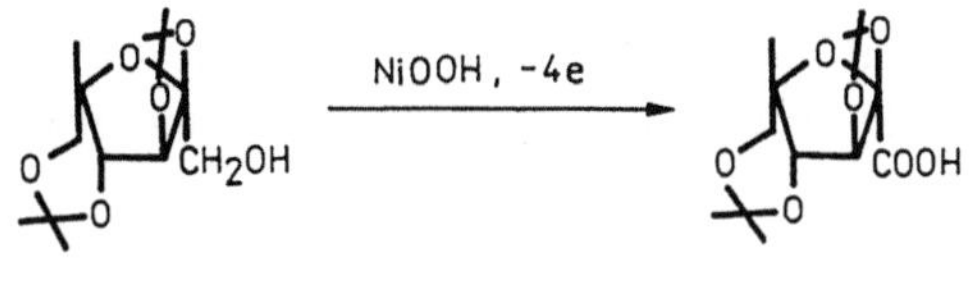

(3-220)

In analogous oxidations of secondary alcohols ketones may be prepared in high yields (3-221).

$$\text{(3-221)}$$

The high chemoselectivity of the oxidation is exhibited particularly in steroid alcohols in which the oxidation occurs most easily in the sterically accessible hydroxyl group at position 3α or 3β; the ease of oxidation of the individual groups in molecules of steroid alcohols was determined in the sequence 3β-OH . . . 3α-OH > 17β-OH ≫ 20β-OH > 11β-OH (3-222).

$$\text{(3-222)}$$

vic-Diols are oxidized in combination with a cleavage giving rise to acids (3-223).

$$\text{(3-223)}$$

1,ω-diols yield alkanedioic acid (3-224).

$$\text{(3-224)}$$

In the oxidation of aldehydes to acids the basicity of the electrolyte and the temperature must be lowered in order to prevent disproportionations and aldol reactions (3-225).

$$\text{(3-225)}$$

57 %.

Primary amines are "dehydrogenated" at a nickel-hydroxide anode and give nitriles in high yields; the oxidation of amines with a shorter chain proceeds already at 5 °C, amines with a longer chain are oxidized at 40 °C (3-226).

$$R-CH_2-NH_2 \xrightarrow{\text{NiOOH, } -e} [\ R-CH=NH\ ] \xrightarrow{\text{NiOOH, } -e} R-C\equiv N$$

$$\text{(3-226)}$$

| $R = C_5H_{11}$ | $H_2O–KOH$ | 72% |
| $C_{10}H_{21}$ | $H_2O–KOH–t–BuOH$ | 91% |

Indirect electrooxidations occur also on anodes the surface of which has been chemically modified by bonding to a suitable, electrochemically regenerated redox system (mediator). By this approach one succeeded in carrying out an indirect anodic oxidation of reduced nicotinamide adenine nucleotide (NADH) to its oxidized form NAD^+ which acts as a coenzyme in a number oxidations catalyzed by enzymes (3-227). The oxidation potential $E_{1/2}$ for the electrochemical oxidation of NADH to NAD^+ at pH $= 7$ is 0.7 V vs SCE.

$$\text{(3-227)}$$

This oxidation, however, can be accomplished at 0.34 V (vs SCE) if *ortho*-dihydroxyaromatic compounds, such as 3,4-dihydroxybenzylamine or 2-(3,4-dihydroxyphenyl)ethylamine (dopamine) are attached to the surface of an anode fabricated from pyrolytic graphite (3-228).

$$(3\text{-}228)$$

The catalytic cycle comprises the direct anodic oxidation of dihydroaromatics (M_{red}) to their ortho-quinoid form (M_{ox}) which oxidizes NADH to NAD^+ and is regenerated again back to M_{red} (3-229).

$$(3\text{-}229)$$

A much larger number of indirect anodic oxidations can be implemented by applying mediators dissolved in the electrolyte. The reduced form of the mediator (M_{red}) is oxidized by a heterogeneous electron transfer to the anode at a lower potential than that of the substrate. In this way the oxidized form of the mediator (M_{ox}) results which in a homogeneous reaction oxidizes the substrate (S) to the product (P) and M_{red}, resulted in this reaction, is transformed by an electrochemical oxidation in the following cycle to M_{ox} again (3-230).

$$(3\text{-}230)$$

If the lifetime of the whole redox system $M_{ox}M_{red}$ is sufficiently long only a catalytic amount of the mediator suffices for the initiation of the reaction.

What conditions must a good mediator fulfill?

1) It must be chemically stable in both oxidation states M_{ox} and M_{red} in order to prevent the decrease of its catalytic activity by the fact that one of the species would be transformed to an electrochemically non-regenerable compound.
2) The heterogeneous electron transfer from the mediator to the anode, as well as the redox reaction with the substrate must be fast enough, in order to

prevent side reactions and not to force the experimenter to construct large-area electrodes.

3) The mediator must not react with the product or with the solvent, or these reactions must be maximally suppressed.

4) Both oxidation states of the mediator must be easily soluble in the electrolyte, except for electrooxidations in a two-phase system.

5) The mediator should be easily separable from the product and its purification before the electrolysis and after it should be easy.

The oxidation of the substrate by the mediator in the solution could essentially proceed in two ways:

a) direct electron transfer from the substrate (SH) to the mediator; such a process is usually called "redox catalysis" or a "homomediator system" (3-231).

$$M \xrightarrow{-e} M^{(\pm)} \xrightarrow[-M]{SH} SH^{(\pm)} \xrightarrow{-H^{(+)}} S^{\bullet} \longrightarrow \longrightarrow P \qquad (3\text{-}231)$$

b) chemical oxidation of the substrate by the oxidized form of the mediator, e.g. by an abstraction of hydrogen or of the hydride ion; such a process is usually called "chemical catalysis with electrochemical regeneration" or also "hetero-mediator" or "chemomediator" system (3-232).

$$M \xrightarrow{-e} M^{(\pm)} \xrightarrow[-MH^+]{SH} S^{\bullet} \longrightarrow \longrightarrow P \qquad (3\text{-}232)$$

In case of chemomediators as electrochemically regenerable redox systems, aromatic amines covering the potential range from 0.76 V to 2.0 V (vs SCE) are most frequently applied. Such a potential range is sufficiently broad to enable the choice of the most suitable mediator (3-233) for a given oxidation.

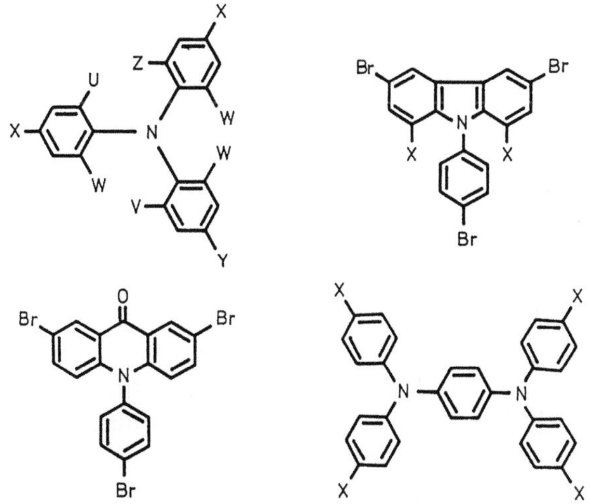

$$(3\text{-}233)$$

X,Y,V,U,W,Z = H, Br, CH_3, CN, CF_3, NO_2

Table 3.3. Standard potentials of some Triarylamines

Name	E^0, (V) vs NHE
N,N,N',N'-tetraphenyl-1,4-benzenediamine	0.75
N,N,N',N'-tetratris(4-bormophenyl)-1,4-benzenediamine	0.93
tris(4-bromophenyl)amine	1.30
tris(4-trifluoromethylphenyl)amine	1.60
tris(4-cyanophenyl)amine	1.68
tris(2,4-dibromophenyl)amine	1.74
tris(2,4,6-tribromophenyl)amine	1.96

Selected amines and their standard potentials are shown in Table 3.3.

The triarylamines are anodically oxidized and yield the corresponding cation radicals, whose aromatic nucleus can be attacked by nucleophilic solvents (acetonitrile, methanol) present in the electrolyte. For this reason their stability is increased by substitution of the phenyl groups by bromine atoms or other substituents (CH_3, CF_3, CN, NO_2, . . .) at positions ortho or para. One of the most stable amines is e.g. tris(2,4-dibromphenyl)amine in which not less than 2500 catalytic cycles were found without any irreversible transformations.

In comparison with direct electrooxidations, when applying anodically generated triarylammonium cation radicals the potential of the working electrode may be lowered by 0.6 to 1.0 V. Thus the direct anodic oxidation of carboxylate ions proceeds at potentials higher than 1.8 V vs SCE. However, in presence of tris(4-bromophenyl)amine such a reaction can be performed at potentials near to the equilibrium potential of the given reaction (1.05 V vs SCE), (3-234).

$$R-C-O^{(-)} \longrightarrow R-C-O^{\bullet}$$

with $Ar_3N^{+\bullet} / Ar_3N$ couple, $-e$ (1.05 V vs SCE)

$$R-COO^{\bullet} \longrightarrow R^{\bullet} + CO_2 \qquad Ar = 4\text{-bromophenyl}$$

$$2R^{\bullet} \longrightarrow R-R \tag{3-234}$$

Indirect anodic oxidations by triarylammonium mediators proceed under very mild conditions and with a very high chemoselectivity which can be "tuned up" by the choice of the amine with a suitable oxidation potential. For this reason these mediators are especially useful, particularly in the oxidative removal of protecting groups. For example, they become a part of the strategy in the transformation of 3-hydroxymethyl-4-heptanol to 3-bromomethyl-4-heptanol which cannot be accomplished by any direct substitution reaction. Chemoselectively, however, it is possible to protect the hydroxyl groups successively by an anisyl and then by a benzyl group. From the resulting diether one can selectively remove only the anisyl group by an anodic oxidation in presence of tris(4-bromophenyl)amine as mediator. The resulting alcohol is transformed to the corresponding bromide in the following step and the remaining benzyl group is then removed by the oxidation with the help of anodically generated tris(2,4-dibromophenyl)ammonium cation radicals (3-235).

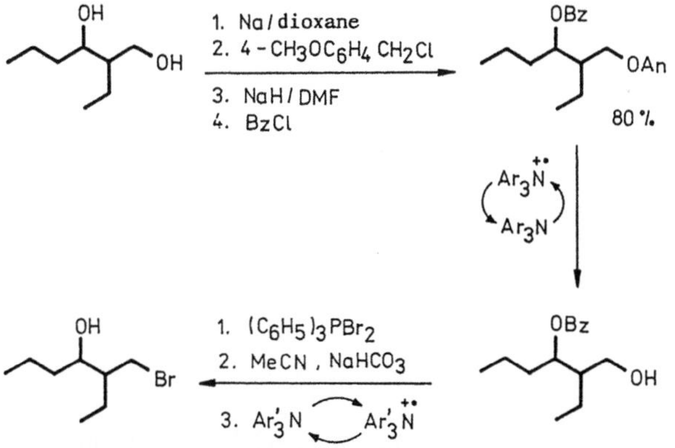

$$(3\text{-}235)$$

Ar = 4-bromophenyl, Ar' = 2,4-dibromophenyl, Bz = benzyl, An = 4-methoxy-benzyl

In both cases the protective groups are transformed to the corresponding aldehydes–anisaldehyde and benzaldehyde. In preparative chemistry this reaction is used in the aimed preparation of aromatic aldehydes from the corresponding alcohols and ethers of the benzyl type (3-236).

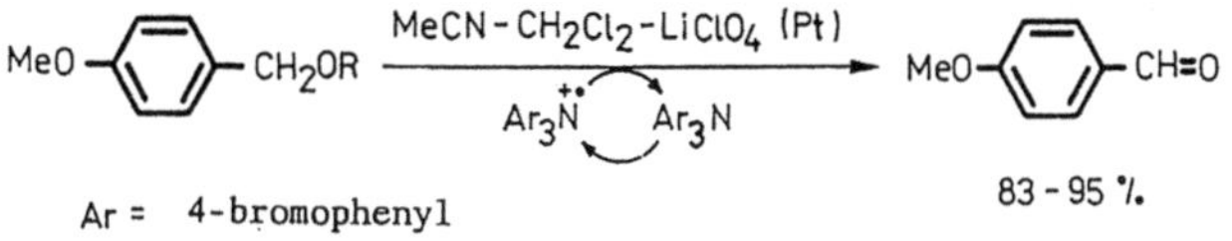

$$(3\text{-}236)$$

The nonactivated hydroxyl groups resist the oxidative influence of ammonium radical cations up to potential values in the vicinity of 1.6 V (vs NHE). The very mild conditions of the above reactions enable hence e.g. a selective liberation of carbonyl compounds from 1,3-dithians or 1,3-dithiolans without attacking the double bonds or hydroxyl groups (3-237) already present in the system.

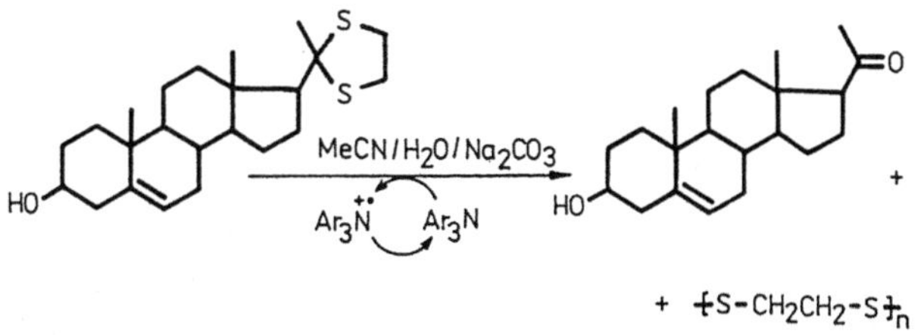

$$(3\text{-}237)$$

The high selectivity of the indirect oxidation under the influence of anodically generated trisarylammonium radicals is evident even in the removal of an anisyl group from the molecule of an unsaturated benzylether in which the difference in the potentials of tris electrophores – the double bond and the anisyl group – amounts to 100 mV (3-238).

$$C_6H_5-CH=CH-(CH_2)_2-O-CH_2-\!\!\left\langle\!\!\bigcirc\!\!\right\rangle\!\!-OMe$$

$E_{pa} = 1,95\ V$ | $E_{pa} = 1,85\ V$

direct oxidation | $Ar_3\overset{+\bullet}{N}\quad Ar_3N$

anode (1,90 V) | anode (1,3 V)

mixture of products $C_6H_5-CH=CH-(CH_2)_2-OH$ (3-238)

80 %

Ar = 4-bromophenyl

By indirect electrolysis in the presence of tris(2,4-dibromophenyl)amine as mediator alkyl- or arylthio groups in thioesters can be activated: this facilitates their hydrolysis to acids; by oxidation of the sulfur atom these groups are transformed to "good leaving" groups (3-239).

$$R-\overset{\underset{\|}{O}}{C}-S-R' \xrightarrow[\;Ar_3\overset{+\bullet}{N}\quad Ar_3N\;]{MeCN\,(0,3\%\ H_2O)\,,\ NaHCO_3/LiClO_4} \left[R-\overset{\underset{\|}{O}}{C}-\overset{+\bullet}{S}-R'\right] \longrightarrow$$

Ar = 2,4-dibromophenyl

$$\xrightarrow[-H^+]{H_2O} R-\overset{\underset{\|}{O}}{C}-OH\ +\ R'SSR' \qquad (3\text{-}239)$$

The reaction was applied for the synthesis of the otherwise only hardly accessible *trans*-cyclopentanolactone from the corresponding *S*-phenyl-*trans*-(2-hydroxycyclopentyl)-thioacetate (3-240).

$$\qquad\qquad \xrightarrow[Ar_3\overset{+\bullet}{N}\quad Ar_3N]{MeCN} \qquad\qquad +\ (C_6H_5S)_2 \qquad (3\text{-}240)$$

52 %

The degree of electrooxidation of methylaromatics depends on the way how the operation is performed. By a direct oxidation they give rise to dimethyl-acetals of aldehydes, in the presence of tris(2,4-dibromophenyl)amine they are oxidized at lower potentials up to the oxidation degree of acids; in neutral media methyl benzoates result, in alkaline solutions trimethylorthobenzoates were detected (3-241).

$$(3\text{-}241)$$

There are not many organic compounds which satisfy the requirements concerning the stability of all oxidation states through which they pass as mediators. In addition to the above-mentioned ortho-quinones and tris-aryl-amines this group also comprised N-hydroxyphthalimide, 2,2,6,6-tetramethyl-piperidyloxide (TEMPO) and 2,3-dichloro-5,6-dicyano-1,4-benzoquinone (DDQ), (3-242).

$$(3\text{-}242)$$

N-Hydroxyphthalimide is anodically oxidized to a phthalimide N-oxylo radical which splits off a hydrogen atom (H·) from the organic substrate. The regeneration of the active radical proceeds at the anode in presence of a base (pyridine, 2,6-lutidine). It is used in the oxidation of secondary alcohols to ketones (3-243).

$$(3\text{-}243)$$

On the other hand, by the anodic oxidation of piperidyl oxide (TEMPO) in the presence of a base an oxoammonium ion is formed which oxidizes the organic substrate in situ in such a way that it splits off a hydride ion. The resulting hydroxypiperidine reacts with the oxoammonium ion again under the formation of piperidyl oxide which in the following cycle is reoxidized to the oxoammonium ion. It can be used again for a selective oxidation of primary alcohols to aldehydes, even in the presence of secondary hydroxyl groups. The oxidation is performed at $-60\,^{\circ}C$ in the system acetonitrile-LiClO$_4$-2,6-lutidine with a platinum anode (3-244).

$$\tag{3-244}$$

The largest group of mediators for indirect anodic oxidations comprises metal ions, particularly the ions of transition metals and a number of their coordination compounds. The electrochemical regeneration of many oxidants has been described such as Cr(VI), Ce(IV), Mn(III), V(V), Fe(III), Hg(II), Pd(II), Te(III), Ag(II), Os(VIII), Fe(CN)$_6^{3-}$, Ru(IV)-complexes, RuO$_4$. These oxidants are transformed from their reduced form M_{red} to M_{ox} by electrooxidation at anodes such as Pb, PbO$_2$, alloys of Pb with metals (Ag, Sb), as well as Pt and Pt(Ti). By a suitable choice of the mediator and by optimization of the reaction conditions a high selectivity of oxidation reactions is achieved and various types of compounds are prepared. Thus, for example, for the "four-electron" oxidation of methylaromatics to the corresponding aldehydes the redox systems Mn(III)/Mn(II) and Ce(IV)/Ce(III) are most frequently used as "oxygen transfer" reagents (3-245).

$$\tag{3-245}$$

R = H, OMe

The technique proper of performing an indirect anodic oxidation is determined by the kind of separation of both processes – the electrooxidation of the mediator from M_{red} to M_{ox} and the chemical oxidation of the organic substrate by the mediator M_{ox}. Both processes may proceed jointly in a simple electrolytic cell, in the so-called "in cell" method (3-246) or can be separated from each

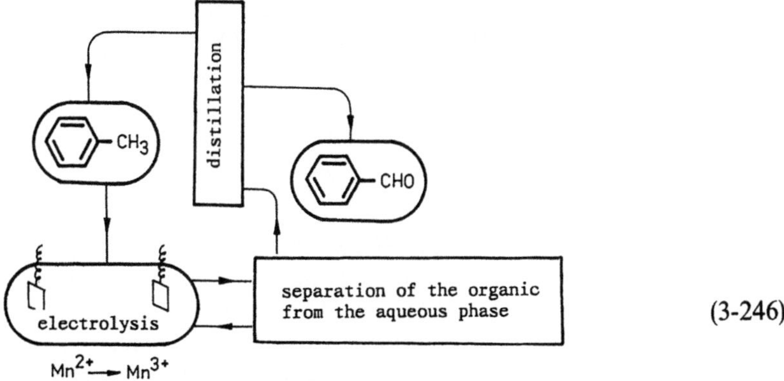

(3-246)

other, in the "ex-cell" method (3-247).

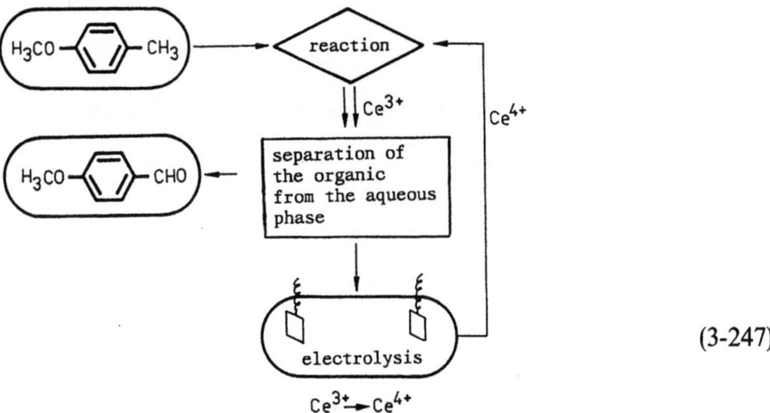

(3-247)

The "six-electron" oxidation of methylaromatics to acids is carried out with Cr(VI)/Cr(III) redox systems and it has been used for example, in the synthesis of sacharine (3-248).

(3-248)

or is performed by electrochemically generated Ru(IV) chelates in the presence of phosphate or borate buffers (3-249).

(3-249)

tpy = terpyridyl; bpy = bipyridyl

The "two-electron" selective acetoxylation of methylaromatics is reached with anodically generated cobalt(III) acetate (3-250).

$$\text{(3-250)}$$

Electrochemically generated Mn(III) acetate is the source of carboxymethyl radicals resulting in the oxidation of the acetate ion in the coordination sphere of Mn(III). The radicals thus formed may be added on to the double bonds of unsaturated compounds; the addition to butadiene is made use of for the synthesis of intermediates required in the production of sorbic acid. The oxidation of the resulting "radical adduct" to the corresponding carbocation can be also considered an indirect anodic oxidation by the electrochemically generated manganese(III) acetate (3-251).

anode : $HOAc + Mn(OAc)_2 \longrightarrow Mn(OAc)_3 + H^+ + e$

$Mn(OAc)_3 + HOAc \longrightarrow Mn(OAc)_2 + HOAc + \dot{C}H_2COOH$

$$\text{(3-251)}$$

The group of indirect electrochemical oxidations of aromatic compounds can also include the oxidation of benzene to phenol in a reaction with the so-called "Fenton reagent" generated in situ. The Fenton reagent is formed, however, at the cathode by the reduction of ferric to ferrous salts and by the reduction of oxygen to hydrogen peroxide. An addition of Cu(II) salts facilitates the reaction because Cu^{2+} ions are better oxidants than the Fe^{3+} ions (3-252).

$$\text{cathode:} \quad Fe^{3+} + e \longrightarrow Fe^{2+}$$

$$O_2 + 2e + 2H^{(+)} \longrightarrow H_2O_2$$

$$Fe^{2+} + H_2O_2 \longrightarrow Fe^{3+} + {}^{\bullet}OH + OH^{(-)}$$

$$HO^{\bullet} + \text{[benzene ring]} \longrightarrow \text{[cyclohexadienyl radical]}-OH \xrightarrow{Cu^{2+}} \text{[phenol]}-OH + Cu^{(+)} + H^{(+)}$$

$$Cu^{(+)} + Fe^{3+} \longrightarrow Cu^{2+} + Fe^{2+}$$

$$\overline{\quad O_2 + H^{(+)} + 2e + \text{[benzene]} \longrightarrow \text{[phenol]}-OH + HO^{(-)} \quad} \tag{3-252}$$

Particularly interesting mediators for indirect oxidations are non-metal ions. Most frequently the systems Br^-/Br^+, Cl^-/Cl^+, I^-/I^+, IO^-/IO_4^+, Cl^-/ClO^-, etc. are used, the redox potentials of which (E_o) vary in alkaline media from $+1.09$ V up to $+1.64$ V. The redox systems of such mediators can be generated by the in-cell method. They are applied for overcoming the relatively large potential differences between the redox calyst and the substrate (as much as 2.0 V). Thus a direct anodic oxidation of secondary alcohols proceeds at potentials approaching 2.8 V (vs SCE). However, they can be indirectly oxidized to ketones in an aqueous solution of *tert*-butyl alcohol with potassium iodide by the anodically generated I^+ ion as soon as at 0.6 to 0.8 V (vs SCE) – (3-253).

$$C_6H_{13}-\underset{\underset{OH}{|}}{CH}-CH_3 + I^{(+)} \xrightarrow{-H^+} C_6H_{13}-\underset{\underset{OI}{|}}{CH}-CH_3 \xrightarrow{-H^+} C_6H_{13}-\underset{\underset{O}{||}}{C}-CH_3 + I^{(-)}$$

$$\text{anode} \quad +0{,}6\,V \qquad 99\,\% \tag{3-253}$$

Under analogous conditions the primary alcohols yield alkylalkanoates (3-254).

$$C_8H_{17}-OH \xrightarrow[I^+ \quad I^-]{KI, H_2O} C_7H_{15}-\underset{\underset{O}{||}}{C}-O-C_8H_{17} \qquad 88\,\% \tag{3-254}$$

When using a polymeric mediator – poly(4-vinylpyridine) hydrobromide (PVP.HBr) one can accomplish a selective oxidation of secondary alcohols in the presence of a primary alcoholic group (3-255).

$$\underset{OH}{\text{[CH}_3\text{CH(OH)CH}_2\text{CH}_2\text{OH]}} \xrightarrow[PVP \cdot HBr]{(Pt), MeCN} \text{[CH}_3\text{C(O)CH}_2\text{CH}_2\text{OH]} \tag{3-255}$$

The different course of the direct and of the indirect anodic oxidation can be demonstrated with the methoxylation of lysine methylester N,N'-biscarbamate. The direct anodic oxidation occurs on the electron pair of the nitrogen atom of the ω-amino group which has a lower oxidation potential than the nitrogen atom in the α-position. On the other hand, Cl^+ formed at the anode yields an N-chloro derivative with an α-amino group; by elimination of hydrogen chloride through the action of a base, resulting in situ, an iminoderivative results which by adding methanol gives rise to the α-methoxylated derivative (3-256).

$$(3\text{-}256)$$

Important from the synthetic point of view is the epoxidation of unsaturated compounds in which Br^+ formed at the anode takes part (3-257).

$$(3\text{-}257)$$

The efficiency of the applied bromides decreases in the following sequence: NaBr (91%) > $(C_2H_5)_4$NBr (82%) > KBr (78%) > LiBr (75%) > NH_4Br (13%).

A transient formation of an enamine corresponding to an epoxide is also assumed in the anodic oxidation of a mixture of aldehydes and secondary amines in the presence of potassium iodide in which N,N-dialkylaminomethyl ketones result (3-258).

$$(3\text{-}258)$$

The electrochemical methoxylation of furan in presence of ammonium bromide (cf. page 87) and the oxidation of glucose to calcium gluconate performed in the presence of bromide ions as redox catalyst can be both considered the oldest indirect anodic oxidations: the oxidant proper is calcium hypobromite which is formed by the reaction of bromine resulting at the anode with calcium carbonate present in the reaction mixture (3-259).

$$\text{(3-259)}$$

Both reactions were introduced in industry as well as the oxidation of starch to the corresponding dialdehyde by the periodate regenerated at the anode (3-260).

$$\text{(3-260)}$$

The indirect anodic oxidations of organic substrates can be also performed by applying "double" redox systems. In this way a further lowering of the potential of the working electrode was achieved as compared to a system with a single mediator.

The oxidation of secondary alcohols to ketones is carried out in the presence of octylmethylsulfide as mediator which gives rise to the redox system $R_2'S/R_2'S^+$ having a potential $E = 1.93$ V vs SCE. As far as bromide ions are present in the mixture to be electrolyzed, the redox system Br^-/Br^+ with $E_p = 1.1$ V vs SCE results; at this potential the electrochemical reaction proper proceeds (3-261).

$$\text{(3-261)}$$

A different double mediator system was used for the hydroxylation of a double bond by Os(VIII) compounds (3-262).

$$\text{(3-262)}$$

$$94 - 99\,\%$$

The indirect electrochemical oxidations of organic compounds can be performed in two-phase systems applying phase-transfer catalysis as follows from the oxidation of benzylalcohols by hypobromite ions. The anodic oxidation of bromides in the aqueous phase leads to the formation of bromine which yields the hypobromite ion by disproportionation in water. This is transported into the organic phase as ammonium hypobromite. The bromide ion resulting in the oxidation is transported into the aqueous phase in the same way (3-263).

$$\text{organic phase} \quad C_6H_5-CH_2-OH \xrightarrow{\;BrO^-\;} C_6H_5-CH=O \;+\; Br^-$$

$$\text{aqueous phase} \quad R_4\overset{(+)}{N}\,Br\overset{(-)}{O} \qquad\qquad\qquad R_4\overset{(+)}{N}\,Br^{(-)}$$

$$\text{anode} \quad 2\,Br^- \xrightarrow{\;-2e\;} Br_2 \underset{}{\overset{H_2O}{\rightleftharpoons}} HOBr \;+\; HBr$$

$$\downarrow R_4NBr$$

$$R_4\overset{(+)}{N}\,Br\overset{(-)}{O} \;+\; HBr$$

$$(3\text{-}263)$$

3.4 Indirect Cathodic Reductions [180, 181]

In the case of indirect cathodic reductions the mediators can be also a component of the cathode (electrocatalyzed reductions) or a component of the electrolyte.

The group of electrocatalyzed reductions comprises all reductions by "amalgamated" cathodes. These electrodes are formed by the reduction of metal salts or even of tetralkylammonium salts at mercury electrodes. Reduction with electrogenerated amalgams can be carried out either by the "in cell" or by the "ex-cell" method. The reductive power of amalgams of the individual metals decreases in the sequence $Li > Na > K > Ba > Sr > Ca$. The reductive power of the amalgam can be to a certain degree varied by the choice of the metal, by its concentration or by the choice of the amalgam surface (3-264). Electrochemically prepared amalgams can be used e.g. for regioselective reductions of double bonds,

$$(3\text{-}264)$$

for hydrodimerization of compounds with an activated double bond (3-265)

$$(3\text{-}265)$$

or for partial reductions of aromatic compounds (3-266).

$$(3\text{-}266)$$

In organic synthesis, metals (Zn, Sn) or their ions in a low oxidation state – Ti(III), V(III), Sn(II), Cr(III), Ce(III) are used as selective reductants. If they are applied in stoichiometric amounts, problems arise with their regeneration or with the liquidation of salts containing ions in a higher oxidation state.

In two-phase systems the metals (Zn, Cu, Fe, Sn) can be electrogenerated with a high current yield in powder form and at high current densities ($\geq 0.5\ \mathrm{A\,cm^{-2}}$). The metals obtained in this way are highly reactive and are used for reductions of nitro compounds and for dehalogenation of *vic*-dihalogeno derivatives (3-267).

$$\mathrm{CH_2Cl-CHCl_2} \xrightarrow[\mathrm{Zn\ \ Zn^{2+}}]{} \mathrm{CH_2{=}CCl_2} + 2\,\mathrm{Cl^{(-)}} \qquad (3\text{-}267)$$

84 %

Cr(II) salts are selectively reducing reagents which are widely applied in organic synthesis. Since they are easily oxidized even by atmospheric oxygen their electrogeneration in situ is particularly convenient; in this form they can be used e.g. for the reductive coupling of alkyl- and benzylhalogenides (3-268)

$$(3\text{-}268)$$

77 – 81 %

further for the reduction and also simultaneously for the reductive elimination of *vic*-halogenohydrines (3-269).

$$(3\text{-}269)$$

The electrochemically prepared complex of ethylenediamine and chromium(II) perchlorate in wet dimethylformamide gives an alternative method for the synthesis of desoxynucleotides (3-270).

$$\text{(3-270)}$$

$$R = CH_3CHOC_2H_5 \, , \quad U =$$

Electrochemically generated Ti(III) salts are used for the reduction of aromatic nitro compounds. The reaction was also applied in the production of 4-aminobenzoic acid. Starting from 4-nitrotoluene one can "couple" both electrode processes; in the first step 4-nitrobenzoic acid is prepared by indirect electrooxidation and in the following step it is transformed by indirect electro-reduction to 4-aminobenzoic acid (3-271).

$$\text{(3-271)}$$

A very broad spectrum of activity has been found with redox catalysts based on complexes of transition metals. From higher oxidation states they are transformed to their active form Ni(O), Ni(I), Co(I), Sn(O) by cathodic reduction. The redox potentials of such mediators and thus also their selectivity for a given substrate can be "tuned up" by the choice of the central atom and of the ligands.

In the reaction with alkylating and acylating reagents, the active form of the mediator undergoes the so-called oxidative addition. This reaction leads to a change in the polarity ("Umpolung") of the alkylating or acylating agent; they are transformed to "carbanions", linked by a coordination bond to the central atom whose oxidation state has been increased in such a reaction. The complex further undergoes an electrochemically induced reductive elimination in which the active form of the coordination compound is regenerated at the cathode; simultaneously a radical or a carbanion is formed which can react with unsaturated compounds in an addition reaction.

A suitable mediator in such reactions is the vitamin B_{12} (cyanocobalamine) – the oxidized form of the mediator containing Co(III) used in amounts from 1 to 10 mol%. The general scheme of the catalytic reductive activation of an alkyl halogenide is illustrated in (3-272).

$$(3\text{-}272)$$

A particular example may be the intramolecular alkylation of 2- and 3-(ω-bromoalkyl)-2-cyclohexenones by which one may arrive at cyclic and spirocyclic ketones (3-273), (3-274).

$$(3\text{-}273)$$

$$(3\text{-}274)$$

If the electrochemical reductive elimination in which the Co(I) complex is regenerated occurs in dark it takes place at more negative potentials (-1.5 V vs SCE) than the cathodic reduction of the starting Co(III) complex to the Co(I) complex (-0.9 V).

The reduction potential for the reductive elimination can be decreased down to -0.9 V (vs SCE) if the alkyl- or acyl Co(III) complex is activated by light (the so-called photoelectrocatalysis). The catalytic cycle of such a reductive and photochemical activation of acylhalogenides by vitamin B_{12} is shown in the following Scheme (3-275).

$$(3\text{-}275)$$

Table 3.4. Reduction potentials of organic mediators for unidirectional electroreductions

Compound	$- E_{1/2}$ (vs SCE)
perylene	1.67
anthracene	1.96
pyrene	2.09
chrysene	2.25
phenanthrene	2.45
naphthalene	2.50
biphenyl	2.70

The above technique was made use of in the synthesis of natural products, e.g. pheromones, 3-substituted steroids etc. (3-276).

$$(3\text{-}276)$$

Among the organic compounds those radical anions and dianions were tested as mediators for indirect electrochemical reductions which prevalently result from the electrochemical reduction of aromatic compounds. Reduction potentials of selected organic mediators in indirect electroreductions are shown in Table 3.4.

They are in some cases applied to the indirect electrohydrogenolysis of alkyl and aryl halogenides, sulfonates, sulfonamides, sulfides and ammonium derivatives. The reductions occur with a high selectivity and at milder conditions than those observed in the absence of mediators (3-277).

$$(3\text{-}277)$$

In synthetic work they are applied for removing protecting groups:

$$(3\text{-}278)$$

$$Py = pyrene$$

$$Ts = CH_3\text{—}\langle\ \rangle\text{—}SO_2$$

The anion radicals prepared by reduction of aromatic compounds cannot be used as mediators in protic media. For these media only cation radicals are suitable which result e.g. by the cathodic reduction of methyl viologen (MV^{2+} = N,N'-dimethyl-4,4-bipyridinium dihydrochloride) (3-279). They can only be used at potentials not more negative than -1.0 V (vs SCE) and they were actually used to an indirect electroreduction of $NAD^{(+)}$ to NADH:

$$(3\text{-}279)$$

The reductive activity of NADH was proved with the reduction of hydroxy-acetone; in the presence of yeast the reduction occurs stereoselectively and by this "electromicrobial" way (R)-1,2-propanediol (3-280) results:

$$(3\text{-}280)$$

4 Acids and Bases Generated at Electrodes [183]

It can be observed in the electrolysis of any solution that the electrolyte turns more acid in close vicinity to the anode and more alkaline in close vicinity to the cathode although the bulk of the solution as a whole remains neutral. This fact attracts the attention not only of theoreticians but also that of organic chemists who attempt to utilize such locally generated acids and bases in organic synthesis.

4.1 Electrochemically Generated Acids (EGA)

By the electrolysis of certain types of supporting electrolytes in organic solvents, acid media are formed in the vicinity of the platinum anode; this change can be applied for the so-called "functionalizations catalyzed by electrochemically generated acids".

The course of such a reaction and the selectivity of the resulting products depends on the character and on the strength of the electrogenerated acid; the latter is given particularly by the combination of the supporting electrolyte and the solvent used. The different pH values in the vicinity of a platinum anode after electrolysis explains e.g. the different representation of products resulting in the anodic oxidation of hexamethylbenzene which is performed in an undivided cell in moist acetonitrile containing lithium perchlorate or tetrabutylammonium tetrafluoroborate (4-1).

$$\text{(hexamethylbenzene)} \xrightarrow{CH_3CN - H_2O - (Pt)} \text{(}CH_2OH\text{ product)} + \text{(}CH_2NHCOCH_3\text{ product)} \tag{4-1}$$

	CH_2OH	$CH_2NHCOCH_3$
$LiClO_4$	5 %	95 %
$(C_4H_9)_4N\,BF_4$	95 %	5 %

By measuring the pH value in the surroundings of the anode immediately after the electrolysis, a lower value (2.19) was found for the system acetonitrile-$LiClO_4$ than that (3.93) for the system acetonitrile-$N(C_4H_9)_4BF_4$. Also the fact that the hydroxymethyl derivative is spontaneously transformed to the acetamido derivative by the electrolysis in presence of $LiClO_4$, points to the catalytic effect of the electrogenerated perchloric acid on the above reaction which is not a redox process. The acid transforms the hydroxymethyl derivative back to the carbocation which reacts with acetonitrile giving rise to the acetamido derivative (4-2).

$$(4\text{-}2)$$

A further example may be represented by the rearrangement of oxirane to a ketone which can be effected in an undivided cell by the influence of an acid generated in the vicinity of a platinum anode (4-3). The existence of this acid and its catalytic effect is confirmed by the following circumstances:

a) the reaction proceeds with a high conversion even after applying a catalytic amount of charge $(0.01–0.1\ F\,mol^{-1})$; it does not occur, however, without passing a charge through the solution.
b) in a divided cell, the reaction proceeds only in the anodic compartment, never in the cathodic one.
c) the presence of an equivalent amount of the base (e.g. pyridine) makes the transformation impossible;
d) the reaction also takes place in the case when oxirane is added into a preelectrolyzed mixture of the supporting electrolyte and the solvent (4-3).

$$(4\text{-}3)$$

electrolyte/solvent	charge $(F\cdot mol^{-1})$	yield (%)
$LiClO_4/CH_2Cl_2$	0.06	91
$LiClO_4/THF$	0.03	86
$(C_2H_5)_4NClO_4/CH_2Cl_2$	0.50	87
$LiBF_4/THF$	0.90	62
$(C_2H_5)_4NBr/CH_2Cl_2$	0.30	0
$CF_3COOLi/CH_2Cl_2/THF$	4.50	0

The acidity of the electrogenerated acid is particularly affected by the combination of the supporting electrolyte (both ions play a role) and of the solvent. The high acidity of the acid resulting at the anode in electrolysis of a solution of lithium perchlorate in CH_2Cl_2 (cf. Fig. 4.1a) is explained by the fact that metallic lithium is deposited at the cathode which in absence of a proton source (i.e. of a protic solvent) cannot form the corresponding base and passes into the bulk of the solution. On the other hand, in close vicinity of the anode "naked" perchlorate ions accumulate which do not possess the corresponding "counter-ions" (the "unbuffered" perchlorate); the latter react with traces of water present in the solvent and thus form perchloric acid jointly with hydroxyl anions which are further anodically oxidized to oxygen or to hydrogen peroxide (4-4).

$$ClO_4^{(-)} + H_2O \;\rightleftharpoons\; HClO_4 + HO^{(-)} \tag{4-4}$$

This reaction renders the whole electrolytic system acidic and, particularly in the neighbourhood of the anode, a strongly acidic reaction zone is formed where a number of acid catalysed reactions may proceed spontaneously.

Weaker EGA are obtained by the electrolysis of a system such as R_4NClO_4 in CH_2Cl_2 because the trialkylamine resulting by the cathodic reaction of tetraalkylammonium ions neutralizes the acid which is formed at the anode.

On the other hand, in the electrolysis of lithium perchlorate in methanol (cf. Fig. 4.1b) lithium deposited on the cathode reacts with methanol, hydrogen is evolved and the resulting lithium methoxide neutralizes the acid formed at the anode. As far as in these systems acid catalyzed reactions occur they proceed in close vicinity of the anode surface during electrolysis and for accomplishing them an excess charge is usually required. Tetraalkylammonium bromides and lithium trifluoroacetate do not represent very suitable electrolytes for generating acids at the anode since their anions are first oxidized on account of water oxidation or, respectively, on account of the oxidation of hydroxyl anions. This latter reaction is necessary for the formation of the acid. The combinations of supporting electrolytes and solvents applied in the anodic generation of acids of different strength are shown in Table 4.1.

The anodically generated acids are broadly exploited in organic synthesis for a number of acid catalysed reactions. Particularly made use of is the fact that the acid in the electrolysed mixture can be only locally active and its strength can be

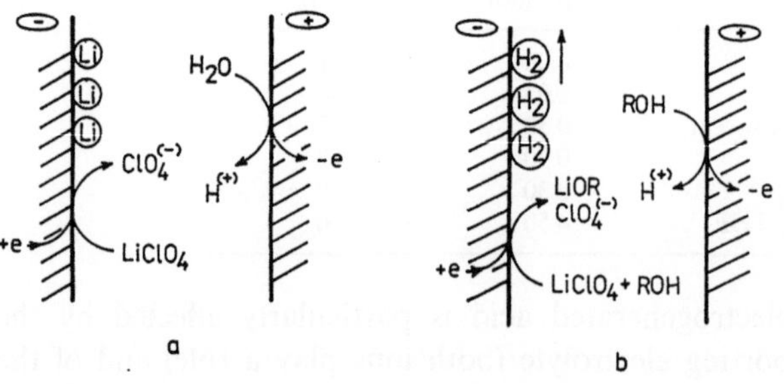

Fig. 4.1. Schematical representation: the formation of electrogenerated acids; **a** – in aprotic solvents, **b** – in protic solvents

Table 4.1. Combinations of supporting electrolytes and solvents for the anodic generation of acids

	Electrolyte	Solvent
	$MClO_4$ (M = Li, Na, Mg)	CH_2Cl_2, CH_2ClCH_2Cl
increasing acidity	MBF_4 (M = Li, Na)	tetrahydrofuran
	R_4NClO_4	ethylacetate
	R_4NBF_4	acetone
	$(C_2H_5)_4NOTs$	acetonitrile
	$(C_2H_5)_4NBr$	methanol

"tuned up" by the suitable choice of the supporting electrolyte and of the solvent. From a number of reactions one can quote e.g. acid catalysed nucleophilic substitutions of alkoxy groups in acetal (4-5):

$$R-CH(OCH_3)_2 \xrightarrow[Me_3SiNu]{LiClO_4 - Bu_4NClO_4 - CH_2Cl_2} R-CH(Nu)(OCH_3)$$

$$Nu = H, CN, CH_2{=}CH-CH_2 \tag{4-5}$$

and crossed aldol condensations of enolethers with acetal (4-6).

$$\tag{4-6}$$

By a suitable combination of the supporting electrolyte and the solvent, also perhaps the duration of the electrolysis, the tetrahydropyranyl group for protecting hydroxylic groups can be introduced, but it can also be selectively removed (4-7).

$$\tag{4-7}$$

In anodic oxidations one must bear in mind that parallel with the substrate oxidation required also the formation of EGA proceeds at the anode. Therefore one must very carefully consider the choice of a suitable electrolyte and its combination with the solvent, in particular as regards the preparation and the isolation of oxidation products sensitive toward acids. On the other hand, this fact can be intentionally made use of. The EGA can catalyze the transformation of a primarily formed product of anodic oxidation to a new intermediate which in the following step undergoes a subsequent anodic oxidation. This sequence of reactions is assumed in the "four-electron" oxidation of methyl-5-acetyl-2-furan-carboxylate, the primary product of the "two electron" oxidation is transformed to a hemiacetal in a reaction catalysed by the electrogenerated acid which is then further oxidized as a *vic*-hydroxy-alkoxy derivative with a simultaneous cleavage of the dihydrofuran cycle (4-8).

$$\tag{4-8}$$

4.2 Electrochemically Generated Bases (EGB)

The cathodic reduction of organic compounds leads to anion radicals and anions as radical intermediates which can behave as bases and/or nucleophiles and/or as reducing agents. The compounds which by the reduction yield intermediates which behave as bases, are called probases (PB) and the electrochemically formed bases are then called electrogenerated bases (EGB).

The electrochemical technique makes it possible to generate bases in situ in aprotic solvents; by the choice of a suitable PB the strength of the EGB can be controlled and by the amount of charge passed through the solution, also, the concentration of EGB.

Several requirements (4-9) must be fulfilled in order to have a PB which by cathodic reduction yields a good EGB (PB$^-$), being able to deprotonate e.g. a weak C-acid (HA):

$$\tag{4-9}$$

a) The reduction potential PB(E_1) must be lower than those of the other solution components of the electrolyte, e.g. of the C-acids (E_2) and of the products of the subsequent reactions ($E_3, \ldots$);
b) EGB should be a strong base but a weak nucleophile and a weak reductant;

c) The protonated form of EGB (PBH) should be easily transformable back to
PB, in particular in the cases where PB is not readily available or expensive.

The organic compounds utilized as PB can be divided into two groups. The
first group comprises the compounds in which after the reduction (after the
formation of EGB) a fission of covalent bonds occurs. Thus by the electrochemical reduction of cyanomethyltriphenylphosphonium salts triphenylphosphine
and the cyanomethyl carbanion are formed; the latter plays the role of EGB by
splitting off a proton from the starting phosphonium salt and transforming it to
the corresponding triphenylphosphoniomethylylide (4-10).

$$(C_6H_5)_3\overset{(+)}{P}-CH_2CN \xrightarrow{2e} (C_6H_5)_3P \ + \ \bar{C}H_2-CN \qquad CH_3-CN$$

$$(C_6H_5)_3\overset{(+)}{P}-CH_2CN \qquad (C_6H_5)_3\overset{(+)}{P}-\overset{(-)}{\bar{C}}HCN$$

$$(4\text{-}10)$$

In the presence of carbonyl compounds the ylides prepared in this way react
in the sense of the Wittig reaction and form the corresponding alkenes (4-11).

$$(C_6H_5)_3\overset{(+)}{P}-CH_2COOMe \ \ Cl^{(-)} \xrightarrow{(C)} \left[(C_6H_5)_3P=CH-COOMe\right] \longrightarrow$$

$$\xrightarrow{R-CH=O} R-CH=CH-COOMe \qquad (4\text{-}11)$$

$$R = C_6H_5 \ (75\%); \ C_3H_7 \ (50\%)$$

In tetrachloromethane a cleavage of the C–Cl bond occurs during the
cathodic reduction, giving rise to a trichloromethyl carbanion which reacts with
dimethyl-4-bromobutyl malonate simultaneously as a base (it deprotonates the
C-acid) and as a nucleophile (it substitutes the C–Br bond), (4-12).

$$CCl_4 \xrightarrow[-Cl^{(-)}]{2e} Cl_3\overset{(-)}{C} \qquad (4\text{-}12)$$

The cathodic reduction of 2-pyrrolidone takes place at considerably negative potentials (− 2.4 V vs SCE). This is why the corresponding amidic anion
cannot be generated in situ. As a rule, it is prepared in advance, by "preelectrolysis" at − 78 °C in dimethylformamide (4-13).

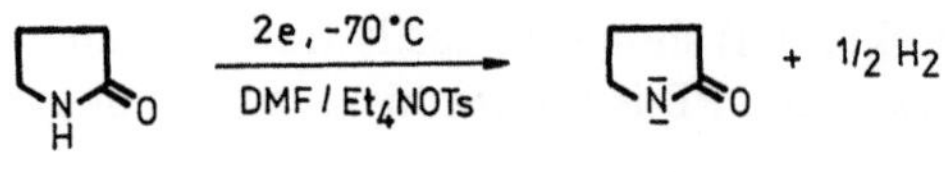

$$\text{(4-13)}$$

It can be used as a base for the deprotonation of C-acids, e.g. in the Dieckman condensation (4-14)

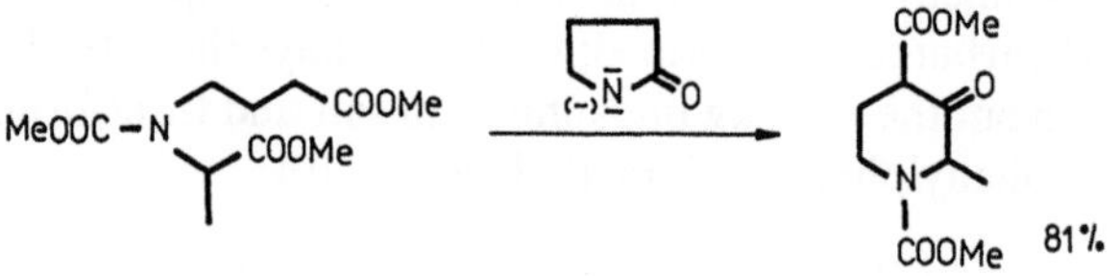

$$\text{(4-14)}$$

or as a nucleophile in the synthesis of α-amino acids (4-15).

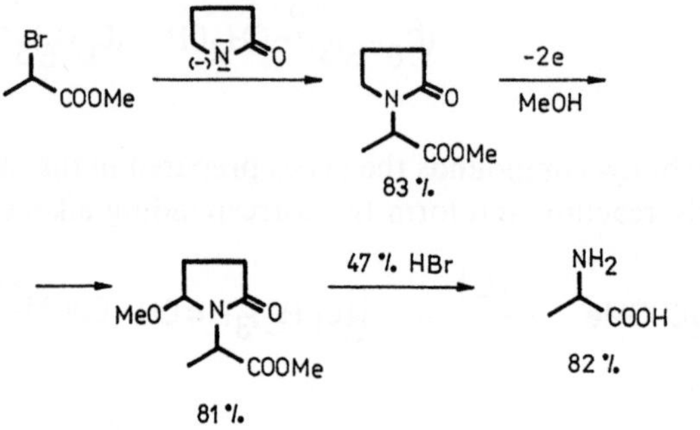

$$\text{(4-15)}$$

The other group of PB is represented by compounds in which no covalent bonds are split in the formation of EGB. The compounds include particularly the aromatic azo derivatives and alkenes with an activated double bond. The azobenzene proper reduces at the cathode at -0.9 V (vs Ag/AgCl) and in its reduced form it is a sufficiently strong base to be used for the deprotonation of C-acids in the Wittig reaction (4-16)

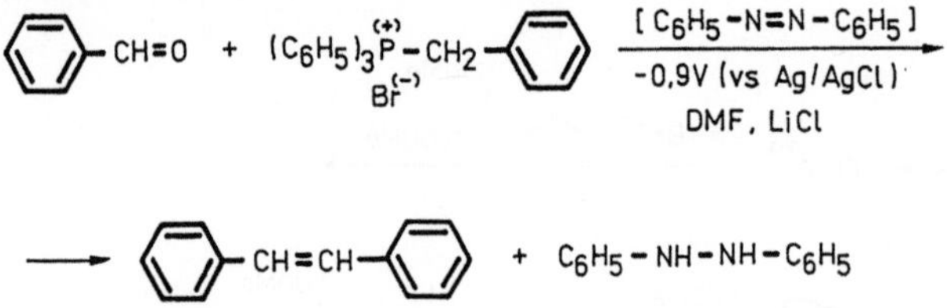

$$\text{(4-16)}$$

or in the intramolecular alkylation of malonic acid esters (4-17).

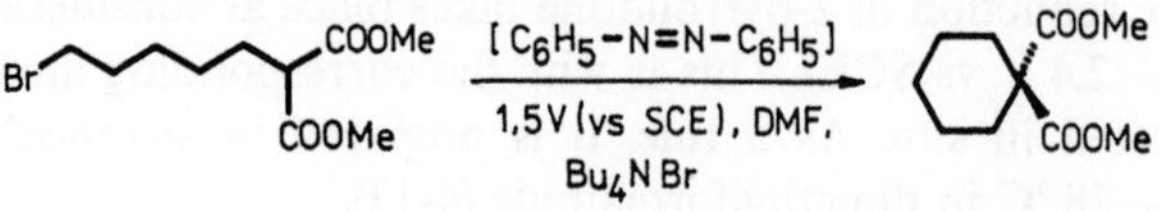

$$\text{(4-17)}$$

The same advantages as in azobenzene (a low reduction potential, a relatively high basicity, easy processing of the reaction mixture) are also exhibited by its alkylated derivatives, such as 2,2'-di-*tert*-butylazobenzene (DTAB); the anion radicals or dianions resulting by their reduction are more basic than the intermediates in the reduction of azobenzene because the *ortho* substituents make the planarity impossible which is necessary for the charge delocalization. Di-*tert*-butylazobenzene was applied for the deprotonation of ethyl phenylacetate in its carboxylation to diethylphenylmalonate (4-18).

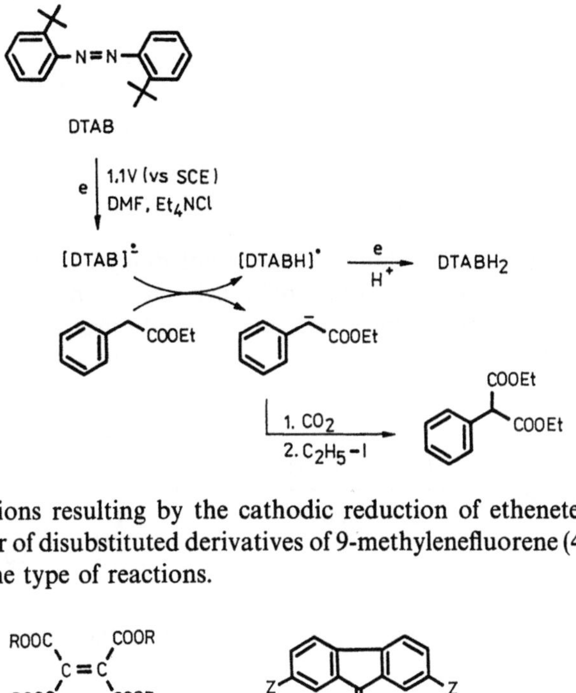

$$(4\text{-}18)$$

The anions resulting by the cathodic reduction of ethenetetracarboxylic acid esters or of disubstituted derivatives of 9-methylenefluorene (4-19) can be used in the same type of reactions.

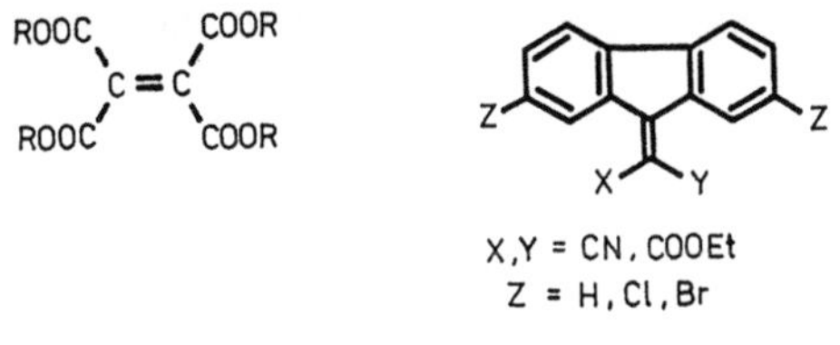

$$(4\text{-}19)$$

An "inorganic" PB is dioxygen by the cathodic reduction of which the superoxide anion radical $O_2^{(-)\cdot}$ is generated. The reduction is performed in a simple H-cell at a mercury cathode in such a way that at $-1.0\,\mathrm{V}$ (vs SCE) oxygen is induced into a solution of a tetraalkylammonium salt in dry acetonitrile or dimethylformamide. The superoxide radical anion alone is a relatively weak base; pK_a of its conjugated acid is 4.69. In the presence of an acid (HA), however, superoxide disproportionates to oxygen and to the hydroperoxide anion. This disproportionation increases its base properties to such a degree that it can perform the deprotonation of acids the pK_a of which reach values approaching 24 (4-20).

$$O_2 \xrightarrow{\;e\;} O_2^{\cdot -}$$

$$O_2^{\cdot -} + HA \rightleftharpoons O_2 + HOO^- + A^- \tag{4-20}$$

The carbanions resulting by deprotonation of C-acids under the influence of the superoxide anion radical react further with an excess of dioxygen. The resulting alkylperoxide anions are transformed to alkoxide ions by the reaction with carbanions; the alkoxide ions act as bases again and deprotonate the C-acid under formation of hydroxy derivatives and new carbanions (4-21).

$$R-H \xrightarrow[-O_2,\,-HOO^-]{2\,O_2^{\cdot -}} R^{(-)} \xrightarrow{O_2} R-O-O^{(-)} \xrightarrow{R^{(-)}} 2\,RO^{(-)} \longrightarrow$$

$$\xrightarrow{2\,RH} 2\,R^{(-)} + 2\,ROH \tag{4-21}$$

By the combination of the superoxide anion radical with dioxygen a new reagent is obtained – the "superoxide-dioxygen reagent" which can be conveniently used in synthesis. The esters of alkylmalonic acid can be transformed to the corresponding α-hydroxyesters in high current yields with respect to the chain mechanism of the reaction (4-22).

$$CH_3-CH(COOEt)_2 \xrightarrow[0.17\ F\,mol^{-1}]{O_2,\ DMF,\ (Hg)} CH_3 - \underset{\underset{\displaystyle OH}{|}}{\overset{\overset{\displaystyle COOEt}{|}}{C}} - COOEt \qquad 95\ \%. \tag{4-22}$$

Nitriles, sulfones and nitroalkanes are transformed to the corresponding carbonyl compounds (4-23).

$$\underset{R^2}{\overset{R^1}{>}}C\underset{X}{\overset{H}{<}} \xrightarrow{O_2^{\cdot -}} \underset{R^2}{\overset{R^1}{>}}\overset{(-)}{C}-X \xrightarrow{O_2} \underset{R^2}{\overset{R^1}{>}}C\underset{O-O^{(-)}}{\overset{X}{<}} \xrightarrow{R^1R^2\overset{(-)}{C}X}$$

$$\longrightarrow 2\ \underset{R^2}{\overset{R^1}{>}}C\underset{\underline{O}^{(-)}}{\overset{X}{<}} \xrightarrow{-X^{(-)}} R^1-CO-R^2 + 2\,X^{(-)} \tag{4-23}$$

$$X = CN,\ SO_2R,\ NO_2$$

Particularly important from the synthetic point of view is the transformation of nitro compounds to ketones in a high yield (an alternative of the Nef reaction). The reaction was made use of in the elegant modification of the synthesis of 1,4-dicarbonyl compounds; in the first step the nitroate anion is prepared from the nitro compound by preelectrolysis in the presence of azobenzene. The nitroate anion reacts with the later added α,β-unsaturated ester in the sense of the Michael addition. The addition product thus formed – a 4-nitro-ester – reacts then by applying the $O_2^{(-)\cdot}/O_2$ reagent to the corresponding 4-oxo-ester (4-24).

$$\text{(4-24)}$$

References to Chapters 3 and 4

1. Koch VR, Miller LL (1973) J Am Chem Soc 95: 8631
2. Bewick A, Edwards GJ, Jones SR, Mellor VM (1977) J Chem Soc, Perkin Trans I 1831
3. Bertram J, Fleischmann M, Pletscher D (1971) Tetrahedron Lett 349
4. Hembrock A, Schäfer HJ (1985) Angew Chem Int Ed Engl 24: 1055
5. Shono T, Matsumura Y (1975) Bull Chem Soc Japan 48: 2861
6. Laurent E, Tardivel R (1976) Tetrahedron Lett 2779
7. Shono T, Matsumura Y, Nakagawa Y (1974) J Am Chem Soc 96: 3532
8. Yoshida K, Kanbe T, Fuenot T (1977) J Org Chem 42: 2313
9. Bewick A, Mellor JM, Pons BS (1978) J Chem Soc, Chem Commun 738
10. Shono T, Ikeda A, Kimura Y (1971) Tetrahedron Lett 3599
11. Shono T, Ikeda A (1972) J Am Chem Soc 94: 7892
12. Adams C, Jacobsen N, Utley JHP (1978) J Chem Soc, Perkin Trans I 1071
13. Banda FM, Brettle R (1974) J Chem Soc, Perkin Trans I 1907
14. Engels R, Schäfer HJ, Steckhan E (1977) Liebigs Ann Chem 204
15. Steckhan E, Schäfer HJ (1974) Angew Chem 86: 480
16. Shono T, Nishiguchi I, Okawa M (1976) Chem Lett 573
17. Koch D, Schäfer HJ, Steckhan E (1974) Chem Ber 107: 3640
18. Schäfer HJ (1987) Kontakte (Darmstadt) 17
19. Shono T, Okawa M, Nishiguchi I (1975) J Am Chem Soc 97: 6144
20. Shono T, Nishiguchi I, Nitta M (1976) Chem Lett 1319
21. Shono T, Kashimura S (1983) J Org Chem 48: 1939
22. Shono T, Matsumura Y, Hamaguchi H (1978) Bull Chem Soc Jap 51: 2179
23. Koch D, Schäfer HJ (1973) Angew Chem 85: 264
24. Sundholm G (1971) Acta Chem Scand 25: 3188
25. Scholl PC, Lentsch SE, Van de Mark MR (1976) Tetrahedron 32: 303
26. White DA, Coleman JP (1978) J Electrochem Soc 125: 1401
27. Jäger P, Hannebaum H, Nohe H (1978) Chem Ing Techn 50: 787
28. Shono T, Matsumura Y, Hashimoto T, Hibino K, Hamaguchi H, Aoki T (1975) J Am Chem Soc 97: 2546
29. Shono T, Hamaguchi H, Matsumura Y, Yoshida K (1977) Tetrahedron Lett 3625
30. Shono T, Matsumura Y, Imanishi T, Yoshida K (1978) Bull Chem Soc Jap 51: 2179
31. Torii S, Inokuchi T, Oi R (1982) J Org Chem 47: 47
32. Shono T, Matsumura Y (1969) J Am Chem Soc 91: 2803
33. Kirsanova AI, Smirnova MG (1978) Elektrokhimiya 14: 627
34. Magno F, Bontempelli G, Pilloni G (1971) J Elektroanal Chem Interfacial Electrochem 30: 375
35. Nokami J, Fujita Y, Okawara R (1979) Tetrahedron Lett 3659
36. Humffray AA, Houghton DS (1972) Electrochim Acta 17: 1435
37. Uneyama K, Torii S (1972) J Org Chem 37: 367
38. Gourly J, Martigny P. Simonet J, Jeminet G (1981) Tetrahedron 37: 1459
39. Bontempelli G, Magno F, Mazzocchin GA (1973) J Electrochem Chem Interfacial Electrochem 42: 57
40. Torii S, Tanaka H, Ukida M (1978) J Org Chem 43: 3223
41. Shono T, Matsumura Y, Mizoguchi M, Hayashi J (1979) Tetrahedron Lett 3861
42. Plattern M, Steckhan E (1980) Tetrahedron Lett 21: 511
43. Torii S, Uneyama K, Ono M (1980) Tetrahedron Lett 21: 2741
44. Laurent E, Tardivel R (1976) Tetrahedron Lett 2779

45. Vincent F, Tardivel R, Mison P (1976) Tetrahedron 32: 1681 (1976)
46. Miller LL, Hoffmann AK (1977) J Am Chem Soc 89: 593
47. Breslow R, Goodin R (1976) Tetrahedron Lett 2675
48. Weinberg NL (ed), (1975) Techniques of chemistry V, Technique of electroorganic synthesis, part 2. Wiley, New York
49. Torii S, Inokuchi S, Misima S, Kobayasi T (1980) J Org Chem 45: 2731
50. Torii S, Uneyama K, Nakai T, Yasude T (1981) Tetrahedron Lett 22: 2291
51. Portis LC, Klug JT, Mann CK (1974) J Org Chem 39: 3488
52. Barry JE, Finkelstein M, Mayeda EA, Ross SD (1974) J Org Chem 39: 2695
53. Blackham AU, Kwak S, Palmer JL (1975) J Electrochem Soc 122: 1081
54. Feldhues U, Schäfer HJ (1982) Synthesis 145
55. Shono T, Matsumura Y, Inoue K, Ohmizu H, Kashimura S (1982) J Am Chem Soc 104: 5753
56. Eberson L, Malmberg M, Nyberg K (1983) Acta Chem Scand B37: 555
57. Shono T, Hamaguchi H, Matsumura Y (1975) J Am Chem Soc 97: 4264
58. Shono T (1984) Tetrahedron 40: 811
59. Shono T, Matsumura Y, Tsubata K, Sugihara Y, Yamane S, Kanazawa T, Aoki T (1982) J Am Chem Soc 104: 6697
60. Thomas HG, Katzer E (1974) Tetrahedron Lett 887
61. Irie K, Ban J (1981) Heterocycles 15: 201
62. Warning K, Mitzlaff M, Jensen H (1978) Justus Liebigs Ann Chem 1707
63. Eberson L (1969) In: Patai (ed) Chemistry of carboxylic acids and esters. Wiley, London
64. Mirkind LA (1975) Uspechi Chimii 44: 2088
65. Schäfer HJ (1981) Angew Chemie Int Ed Engl 20: 911
66. Seidel W, Schäfer HJ (1980) Chem Ber 113: 3898
67. Klünenberg H, Schäfer HJ (1978) Angew Chemie Int Ed Engl 17: 47
68. Huhtasaari M, Schäfer HJ, Beckung L (1984) Angew Chem Int Ed Engl 23: 980
69. Schäfer HJ, Pistorius R (1972) Angew Chem Int Ed Engl 11: 841
70. Wuts PGM, Sutherland C (1982) Tetrahedron Lett 23: 3987
71. Yoshikawa M, Kamigauchi T, Ikeda Y, Kitagawa I (1981) Chem Pharm Bull 29: 2582
72. Muck DL, Wilson ER (1970) J Electrochem Soc 117: 1358
73. Shono T, Hayashi J, Otomo H, Matsumura Y (1977) Tetrahedron Lett 2667
74. Taguchi T, Takahashi Y, Itoh M, Suzuki A (1974) Chem Lett 1021
75. Brettle R, Seddon D (1970) J Chem Soc (C) 2175
76. Morgat JC, Pallaud R (1965) CR Acad Sci 260: 5579
77. Schäfer H, Al Azrak A (1972) Chem Ber 105: 2398
78. Nyberg K (1971) Acta Chem Scand 25: 534
79. Kerr JB, Jempty TC, Miller LL (1979) J Am Chem Soc 101: 7338
80. Miller LL, Stewart RF, Gillespie JP, Ramachandran V, So YH, Stermitz FR (1978) J Org Chem 43: 580
81. Eberson L, Nyberg K (1966) J Am Chem Soc 88: 1686
82. Blum Z, Cedheim L, Nyberg K, Eberson L (1975) Acta Chem Scand B29: 715
83. So YH, Miller LL (1976) Synthesis 468
84. Hammerich O, Parker VD (1974) J Chem Soc, Chem Commun 245
85. Eberson L, Radner F (1980) Acta Chem Scand B34: 739
86. Eberson L, Oberrauch E (1981) Acta Chem Scand B35: 193
87. Torii S, Tanaka H, Inokuchi T, Nakane S, Akada M, Saito N, Sirakawa T (1982) J Org Chem 47: 1674
88. Ohmori H, Matsumoto A, Masui M (1980) Chem Pharm Bull 28: 1887
89. Rakoutz M, Michelet D, Brossard B, Varagnat J (1978) Tetrahedron Lett 3723
90. Weinberg SM, Epling GA, Comi R, Reitano M (1975) J Org Chem 40: 1356
91. Ronsold K, Kash H (1979) Tetrahedron Lett 4463
92. Nishiguchi I, Hirashima T (1985) J Org Chem 50: 539
93. Parker VD, Burgert BE (1968) Tetrahedron Lett 2411
94. Rieker A, Dreher EL, Giesel H, Khalifa MH (1978) Synthesis 851
95. Manning MJ, Raynolds PW, Swenton JS (1976) J Am Chem Soc 98: 5008
96. Buchanan GL, Raphael RA, Taylor R (1973) J Chem Soc, Perkin Trans I 373
97. Johnson RW, Bendnarski MD, O'Leary BF, Grover ER (1981) Tetrahedron Lett 22: 3715
98. Yoshida K, Nagase S (1979) J Am Chem Soc 101: 4268
99. Andreades S, Zahnow EW (1969) J Am Chem Soc 91: 4181

100. Rozhkov IN, Gambaryan NP, Galpern EG (1976) Tetrahedron Lett 4819
101. Elming N (1960) Adv Org Chem 2: 67
102. Tanaka H, Kobayasi Y, Torii S (1976) J Org Chem 41: 3482
103. Iwasaki T, Nishitani T, Horikawa H (1982) J Org Chem 47: 3799
104. Shono T, Matsumura Y, Yamane S (1981) Tetrahedron Lett 22: 3269
105. Froborg J, Magnusson G, Thoren S (1975) J Org Chem 40: 122
106. Eberson L (1980) Acta Chem Scand B34: 747
107. Pickup PG, Osteryoung RA (1984) J Am Chem Soc 106: 2294
108. Tourillon G, Garnier F (1982) J Electroanal Chem 135: 173
109. Junghans K (1974) Chem Ber 107: 3191
110. Junghans K (1973) Chem Ber 106: 3465
111. Wawzonek S (1971) Synthesis 285
112. Benkeser RA, Mels SJ (1969) J Org Chem 34: 3970
113. Benkeser RA, Tincher CA (1968) J Org Chem 33: 2727
114. Campbell KN, Young EE (1943) J Am Chem Soc 65: 965
115. Wagenknecht JH (1983) J Chem Educ 60: 271
116. La Perriere DM, Willett BC, Carrol WF, Torp EC, Peters DG (1978) J Am Chem Soc 100: 6293
117. Fry AJ, Mitnick M, Reed RG (1970) J Org Chem 35: 1232
118. Pletcher D, Razaq M (1980) J Appl Electrochem 10: 575
119. Nagao M, Akashi NST, Yoshida T (1966) J Am Chem Soc 88: 3447
120. Fry AJ, Moore RH (1968) J Org Chem 33: 1283
121. Jansson R (1984) Chem Eng News 43
122. Lehmkuhl H (1973) Synthesis 377
123. Horner L, Röder H (1988) Chem Ber 101: 4179
124. Horner L, Röder H (1977) Phosphorus 2067
125. Gunawardena NE, Pletcher D (1983) Acta Chem Scand B37: 549
126. Tallec A (1969) Ann Chim (Paris) 4: 67
127. Tomilov AP, Smirnov VA, Kagan EŠ (1981) Elektrochimičeskije sintězi organičeskich preparatov; Rostov University
128. Volke J, Skála V (1972) J Electroanal Chem 36: 393
129. Vlček AA, Volke J, Pospíšil L, Kalvoda R (1986) In: Rossiter BW (ed) Physical methods of chemistry, vol 2, Electrochemical Methods. J Wiley, New York
130. Park K, Pintauro PV, Baizer MM, Nohe K (1985) J Electrochem Soc 132: 1850
131. Horner L, Hönl H (1977) Chem Ber 110: 2036
132. Viscontini M, Bühler H (1966) Helv Chim Acta 49: 2524
133. Pienemann T, Schäfer HJ (1987) Synthesis 1005
134. Karrenbrock F, Schäfer HJ (1978) Tetrahedron Lett 1521
135. Baizer MM, Chruma IL (1972) J Org Chem 37: 1951
136. Shono T, Ohmizu H, Kise N (1982) Tetrahedron Lett 23: 4801
137. Shono T, Kise N, Suzumoto T (1984) J Am Chem Soc 106: 259
138. Klaus H, Schäfer HJ (1985) Tetrahedron Lett 26: 4899
139. Weinberg NL, Hoffmann AK, Reddy RB (1971) Tetrahedron Lett 2271
140. Berube D, Caza J, Kimmerle FM, Lessard J (1975) Can J Chem 53: 3060
141. Buchtiarov AB, Tomilov AP (1980) Usp Chimii 49: 470
142. Shono T, Nishiguchi I, Ohmizu H, Mitani M (1978) J Am Chem Soc 100: 545
143. Christensen L, Iversen PE (1979) Acta Chem Scand B33: 352
144. Iversen PE, Lund H (1974) Acta Chem Scand B28: 827
145. Gottlieb R, Pfeiderer W (1981) Justus Liebigs Ann Chem 1451
146. Shono T, Nishiguchi I, Ohmizu H (1977) Chem Lett 1021
147. Shono T, Nishiguchi I, Ohmizu H (1977) J Am Chem Soc 99: 7396
148. Engels R, Schäfer HJ (1978) Angew Chem Int Ed Engl 17: 460
149. Nugent ST, Baizer MM, Little RD (1982) Tetrahedron Lett 23: 1339
150. Gassman PG, Rasmy OM, Murdock TO, Saito K (1981) J Org Chem 46: 5457
151. Shono T, Yoshida K, Ando K, Usui Y, Hamaguchi H (1978) Tetrahedron Lett 4819
152. Degrand C, Campagnon PL, Belot G, Jacquin D (1980) J Org Chem 45: 1189
153. Beck F (1972) Angew Chem Int Ed Engl 11: 760
154. Kariv E, Cohen BJ, Gileadi E (1972) Tetrahedron 27: 805
155. Powell LA, Wightman RM (1979) J Am Chem Soc 101: 4412
156. Horner L, Skaletz DH (1975) Justus Liebigs Ann Chem 1210
157. Danly DE (1984) J Electrochem Soc 131: 435C

158. Petrovich JP, Anderson JD, Baizer MM (1966) J Org Chem 31: 3897
159. Mandell L, Daley RF, Day RA (1976) J Org Chem 41: 4087
160. Briscoe MW, Chambers RD, Silvester MJ, Drakesmith FG (1988) Tetrahedron Lett 29: 1295
161. Fritz HP, Kornrumpf W (1978) Justus Liebigs Ann Chem 1416
162. Fritz HP, Kornrumpf W (1979) J Electroanal Chem 100: 217
163. Fry AJ, Reed RG (1972) J Am Chem Soc 94: 8475
164. Fuchigami T, Nonaka T, Iwata K (1976) J Chem Soc, Chem Commun 951
165. Casanova J, Rogers HR (1974) J Org Chem 39: 2408
166. Matschiner H, Voigtländer R (1977) Z Chem 17: 102
167. Casanova J, Rogers HR (1974) J Org Chem 39: 3803
168. Shono T, Matsumura Z, Kashimura S (1978) Chem Lett 69
169. Shono T, Matsumura Z, Kashimura S (1980) Tetrahedron Lett 21: 1545
170. Satoh S, Itoh M, Tokuda M (1978) J Chem Soc, Chem Commun 481
171. Rifi MR (1971) J Org Chem 36: 2017
172. Caroll WF, Peters DG (1980) J Am Chem Soc 102: 4127
173. Shono T, Matsumura Y, Tsubata K, Sugihara Y (1982) J Org Chem 47: 3090
174. Mairanovsky VG (1976) Angew Chem Int Ed Engl 15: 281
175. Horner L, Singer RJ (1969) Liebigs Ann Chem 723: 1
176. Stouwe C, Schäfer HJ (1981) Chem Ber 114: 946
177. Sommelhack MS, Heisohn GE (1972) J Am Chem Soc 94: 5139
178. Engels J (1979) Angew Chem Int Ed Engl 18: 148
179. Husstedt U, Schäfer HJ (1979) Synthesis 964
180. Steckhan E (1986) Angew Chem Int Ed Engl 25: 683
181. Steckhan E (1987) In: Topics in Current Chemistry 142: 1, Springer Verlag, Berlin
182. Schäfer HJ (1987) In: Topics in Current Chemistry 142: 101, Springer, Berlin Heidelberg New York
183. Uneyama K (1987) In: Topics in Current Chemistry 142: 167, Springer, Berlin Heidelberg New York

Springer-Verlag and the Environment

We at Springer-Verlag firmly believe that an international science publisher has a special obligation to the environment, and our corporate policies consistently reflect this conviction.

We also expect our business partners – paper mills, printers, packaging manufacturers, etc. – to commit themselves to using environmentally friendly materials and production processes.

The paper in this book is made from low- or no-chlorine pulp and is acid free, in conformance with international standards for paper permanency.